Sustainable Engineering

Sustainable Engineering
Principles and Implementation

Catherine N. Mulligan

Routledge
Taylor & Francis Group

LONDON AND NEW YORK

CRC Press
Taylor & Francis Group
6000 Broken Sound Parkway NW, Suite 300
Boca Raton, FL 33487-2742

First issued in paperback 2020

ISBN-13: 978-1-4987-7458-1 (hbk)
ISBN-13: 978-0-367-65666-9 (pbk)

Library of Congress Cataloging-in-Publication Data

Names: Mulligan, Catherine N., author.
Title: Sustainable engineering : principles and implementation / Catherine Mulligan.
Description: Boca Raton : Taylor & Francis, a CRC title, part of the Taylor & Francis imprint, a member of the Taylor & Francis Group, the academic division of T&F Informa, plc, 2019. | Includes bibliographical references and index.
Identifiers: LCCN 2018043073 | ISBN 9781498774581 (hardback : acid-free paper) | ISBN 9781498774598 (ebook)
Subjects: LCSH: Sustainable engineering. | Conservation of natural resources. | Environmental protection. | Pollution. | Green products.
Classification: LCC TA163 .M85 2018 | DDC 620.0028/6—dc23
LC record available at https://lccn.loc.gov/2018043073

Visit the Taylor & Francis Web site at
http://www.taylorandfrancis.com

and the CRC Press Web site at
http://www.crcpress.com

Contents

Preface

The purpose of *Sustainable Engineering: Principles and Implementation* is to provide a comprehensive overview of the interdisciplinary field of sustainability as it applies to engineering and methods for implementation of sustainable practices. Due to the constraints on resources and the environment, engineers are facing with new challenges. While it is generally believed that the concepts of sustainable development must be supported to protect future generations, in practice execution is easier said than done. Therefore, the focus of this book is to provide both a conceptual understanding and practical skills to apply these principles to engineering design.

The 27 principles in the 1992 Rio Declaration highlight the importance of protection of environmental quality while meeting the needs of population growth. These principles were reinforced in the 2002 World Summit on Sustainable Development (WSSD) in Johannesburg. Many believe that such statements are not practical or scientifically feasible as development incurs the depletion of resources that cannot be sustainable. However, the declarative statements of principles promulgated by WSSD (2002) and the Rio +20 Summit in 2012 and others that have followed show the need to develop the knowledge and tools to address the goals of sustainability. Some of the most recent goals proposed for 2030 aim to:

- Ensure availability and sustainable management of water and sanitation for all
- Ensure access to affordable, reliable, sustainable, and modern energy for all
- Build resilient infrastructure, promote inclusive and sustainable industrialization, and foster innovation
- Make cities and human settlements inclusive, safe, resilient, and sustainable
- Take urgent action to combat climate change and its impacts

These goals are highly relevant for engineers. Furthermore, to address climate change, countries adopted the Paris Agreement at the COP21 in Paris on December 12, 2015. In the agreement, all countries agreed to work to limit global temperature rise to well below 2°C and, given the grave risks, to strive for 1.5°C, which presents a significant challenge for engineers. Engineers design many systems that require materials and energy and produce transportation, buildings, products, and other structures that can have significant impacts on environment, economy, and society. In so doing, they need to work with many professionals to ensure sustainable system designs.

This book concentrates on introducing relevant theory and principles and ethical expectations for engineers. The concepts related to industrial ecology, green engineering, ecodesign, and ecological footprint among others are presented. Frameworks are presented to indicate the challenges and constraints of applying sustainable development principles. They include relevant frameworks and ISO standards and other management practices, which currently exist to provide the context. The content is designed to be up-to-date and multidisciplinary from an engineering point of view to show where and how sustainable development concepts can be incorporated into engineering design. Tools for working toward solutions, the elements of resources, materials, and energy selection, use, and management are discussed. This is followed by evolution of the concepts of environmental, social, and economic indicators as methods of measuring and comparing options. The tools and models that incorporate the indicators are discussed. Finally, case studies and examples from around the world are provided to demonstrate the incorporation of sustainable practices into design. This book is designed to be used by undergraduate college and university students in any engineering program and other programs in which sustainability is taught, in addition to practicing scientists and engineers and all others concerned with sustainability of products, projects, and processes. An overview of the book follows.

Chapter 1 introduces the concepts of sustainable development. Engineers have to put it in terms for engineering practice for infrastructure, process, and project design. There are many challenges for the future including increasing population and pollution, scarcity of water, loss of biodiversity, increased urbanization, increasing energy requirements and resource depletion, and climate change to name a few. Therefore, these challenges are creating an urgent need to implement sustainable engineering practices. First, the concept and short history of sustainable development are introduced. This is followed by a short summary of some challenges faced by engineers and some impacts of human activities on the environment. The concept of sustainable engineering is introduced and some conclusions are drawn based on these concepts.

Chapter 2 describes some recent advances in engineering practice and design that form the basis of sustainable engineering practice. Engineers design many systems that require materials and energy and produce transportation, buildings, products, and other structures that can have significant impacts on environment, economy, and society. Several tools and guidelines have been developed. Analytical methods are required to evaluate and reduce environmental, economic, and social impacts. Some of these concepts are described in this chapter. To determine if progress is being made, goals must be established for measurement of the progress, feedback, and adaptation for continuous improvement. Indicators of some form are used to measure various components. Various frameworks that have been developed to facilitate assessment and monitoring will be discussed. Some of these are introduced including the Global Reporting Initiative (GRI), triple bottom line, and the Natural Step.

Chapter 3 emphasizes the whole life cycle of a product or process. The life cycle includes the choice of the raw materials to the final use and disposal to minimize environmental effects. An improved overall picture of the environmental impact of a product or process can be obtained over its life cycle. Life cycle assessment (LCA) is a tool that was developed by engineers and is used for industrial ecology to develop an understanding of environmental impact throughout a life cycle. It allows extending thinking beyond the process itself. The LCA concept is described and its use to minimize waste and reduce use of resources, to measure sustainability, evaluate more sustainable alternatives, optimize the product or process design, and indicate where improvement in the process can be made. In addition, the concepts of life cycle costing, social life cycle assessment, and life cycle sustainability assessment are introduced.

Chapter 4 introduces guidance and describes standards related to sustainable products and processes. At the international level, the leading non-governmental organization for standards is the International Standards Organization or ISO. It develops various standards as well as the ASTM International and National Institute of Standards and Technology (NIST). The Environmental Protection Agency (EPA) in the United States develops various environmental test methods and regulatory standards for water, soil, wastes, and air. There are few standards on sustainability, particularly as measurability is a requirement.

Chapter 5 elaborates that engineers are employed in many different industries, each of which has its own activities, such as mining, food processing, forestry, manufacturing, energy, and service. These industries are essential to the economy and support of society. However, their activities can conflict with the goals of a sustainable environment such as: (a) the use of renewable and nonrenewable materials, (b) the use of nonrenewable natural energy resources, and (c) the emissions of atmospheric, liquid, and solid wastes that can impact human health and the environment. Sustainable engineering practices should aim to: (a) more efficiently use renewable and nonrenewable natural resources to allow future generations to continue to benefit from these resources and (b) reduce the discharge of waste materials into the environment to minimize the impact on humans and the environment. The discussion in this chapter concerns the role of engineering in providing more sustainable processes in relation to material, energy, and water use in particular.

Chapter 6 describes the pollutants emitted into the atmosphere, water, and soil environment. The pollutants can accumulate in organisms or be transported or transformed in the environment. Some examples include dry cleaning solvents, various oils including lubricating oil, automotive oil, hydraulic oil, fuel oil, and biosolids from wastewater plants, processing wastes from various industries such as pulp and paper deinking sludges, and organic and inorganic aqueous wastes and wastewaters. Soil and water contamination is the result of accidental spills and leaks, generation of chemical waste leachates and sludges from cleaning of equipment, residues left in used containers, and outdated materials. Therefore, adequate methods for storage and disposal are required for avoidance of the contamination. However, once contamination occurs in the soil or water, remediation may be necessary. Thus, the remediation or treatment should be performed in as sustainable a manner as possible. Practices for remediation are discussed in this chapter.

Chapter 7 discusses the selection and use of sustainable development indicators (SDIs). Sustainability incorporates environmental, economic, and social aspects. Numerous organizations involved in the development of SDIs include the European Environment Agency (EEA), United Nations Development Programme (UNDP), The World Bank, World Watch Institute, International Institute of Sustainable Development (IISD), New Economics Foundation (NEF), United Nations Commission for Sustainable Development (UNCSD), World Tourism Organization (WTO) and nationally the Department of Culture Media and Sport (DCMS), and the Department for Environment Transport and the Regions (DETR). In 1997, the Global Reporting Initiative was put forward by the United Nations Environment Programme (UNEP). These indicators have been widely adopted by many corporations. Engineers and scientists have the largest influence at the technological scale, and thus these types of indicators are the most relevant for their use. Therefore, various indicators and tools that employ them for decision-making are discussed in this chapter.

Chapter 8 focuses on the future needs for sustainability that have gained significant attention in the past decades, particularly by engineers, but they are quite vague and difficult to implement. More recently, tools have been developed to assist in integrating sustainability into design, particularly for buildings and infrastructure, but none of these have been adapted universally. This chapter provides examples of sustainable engineering practices and presents the challenges and needs for the future in education and research. Education programs need to be expanded for engineers to include sustainable practices not as a special course but integrated into existing courses. Improved tools are needed. More partnerships are needed between social scientists, physical science practitioners, health and engineering professionals, educational institutions, governance agencies, and society. In summary, technology, education, regulation, and standards are all essential to promote and implement sustainable engineering practices. Interdisciplinary programs are necessary for training engineers in the circular economy. Research is needed to develop innovative solutions to this changing world under the influence of climate change and increasing uncertainty, deteriorating infrastructure, introduction of new chemicals into the environment, centralization and lock-in of technologies, and growing population to name a few. Engineers have an ethical requirement to rise to this challenge and their role in sustainable development has been undervalued; hence, it is critical.

The author finally would like to acknowledge and is highly grateful for the significant contributions of the, too many to mention, research students and colleagues who have helped in the completion of this book. Dr. Nayereh Saborimanesh is particularly acknowledged for her contribution on the evaluation of reclaimed water, in addition to the industrial partners for providing the valuable case studies.

Author

Dr. Catherine N. Mulligan, Eng. obtained her B.Eng. and M.Eng. degrees in chemical engineering and a Ph.D. in geoenvironmental engineering from McGill University. She worked for the Biotechnology Research Institute and then SNC Research Corp. of SNC Lavalin before joining Concordia University in 1999. She currently holds a Concordia Research Chair in Geoenvironmental Sustainability (Tier I) and is a full-time professor in the Department of Building, Civil and Environmental Engineering.

She has authored more than 95 refereed papers in various journals, authored, co-authored, or edited 8 books, holds 3 patents, and has supervised to completion of more than 55 graduate students. She is the founder and director of the Concordia Institute for Water, Energy, and Sustainable Systems. The institute trains students in sustainable development practices and performs research into new systems, technologies, and solutions for sustainability. She is a Fellow of the Canadian Society for Civil Engineering (CSCE) and the Engineering Institute of Canada (EIC), and a winner of the John B. Sterling Medal of the EIC.

1

An Introduction to Sustainable Development and Engineering

1.1 Introduction

Many definitions of sustainable engineering are available. However, to implement the concepts of sustainable development, engineers have to put it in terms of engineering practice for infrastructure, process, and project design. There are many challenges for the future including increasing population and pollution, scarcity of water, loss of biodiversity, increased urbanization, increasing energy requirements and resource depletion, and climate change to name a few (Butchart et al. 2010). Therefore, these challenges are creating an urgent need to implement sustainable engineering practices. First, the concept and short history of sustainable development are introduced. This will be followed by a short summary of some challenges faced by engineers and some impacts of human activities on the environment. The concept of sustainable engineering will be introduced and then some conclusions will be drawn based on the concept.

1.2 Introduction to Sustainable Development

Since the 1960s, environmental quality became a major issue and concern. Sustainable development was not conceptualized until the 1980s when the World Commission on Environment and Development (WCED) in their 1987 Report defined it. Since then, there have been many definitions of sustainable development. However, the WCED definition remains the most accepted. It indicates that sustainable development is "development that meets the needs of the present without compromising the ability of future generations to meet their own needs" (WCED 1987). Therefore, it has been generally understood that environmental aspects are included within social and cultural ones. It has been considered as environmentalism with a human approach (Allenby 2012).

Since the beginning of the industrial revolution, Glasby (2002) indicated that "mankind's occupation of this planet has been markedly unsustainable" and that the WCED concept of sustainable development is not achievable presently due to the depletion of nonrenewable resources, and excessive exploitation of renewable resources. Ainger and Fenner (2014) postulated that "human development must be sustainable within environmental limits: that is, must be able to continue indefinitely within the environmental carrying capacity of our one planet." This statement, similar to that of Yong et al. (2014), is that "all the activities associated with development in support of human needs and aspirations, must not compromise or reduce the chances of future generations to exploit the same resource base to obtain similar or greater levels of yield."

Social and economic developments are essential globally. The UN Human Development Index (HDI) is a measure of these (http://hdr.undp.org/en/content/human-development-index-hdi). Health, education, and gross domestic product are combined into a score of 1.000. The target is 0.8. The scores of some countries are very low (less than 0.4) due to quality of life while developed countries need to reduce their ecological footprint to 1.9 ha/person (worldwide average). The HDI was developed by the United Nations. A long, healthy life, adequate education, and decent living standards are included in the index. These are measured by life expectancy, actual or expected years of schooling, and gross national income, respectively. All scores are normalized and aggregated using a geometric mean. Norway is ranked first with an HDI of 0.944. The United States' HDI is 0.915. Other countries, such as India, have much lower HDI of 0.609 and are ranked 1,309 (UNDP 2016). A general grouping according to development is shown in Figure 1.1.

Figure 1.2 shows the triple bottom line concept often used to incorporate the three aspects of sustainable development (environmental, social, and economical). It was introduced in 1997 by Elkington (1998). It is also referred to as the three pillars of sustainability. Resilience is another aspect that is often considered along with sustainability. It refers to the capacity to change while maintaining its main characteristics. This reflects its origin in ecology where function and structure remain despite disturbances in the system (Walker and Salt 2006). Resilience differs from sustainability in that it focuses more on the process of adapting to change whereas sustainability focuses on outcomes (Redman 2014).

The aspect of future generations, in the definition of sustainable development, is difficult to estimate, as there are many unknowns. However, ethically there must be a responsibility toward future generations. In the past, there have been many incidences of environmental mismanagement. Commoner (1971) has suggested that society has been impacted by short-term planning. Activities must have a minimal impact on the environment. The intent of this book is to provide an understanding and guidance for

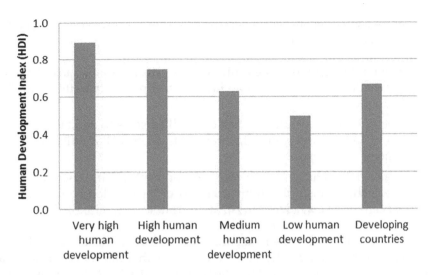

FIGURE 1.1
HDI according to human development group. (Data from UNDP [2016].)

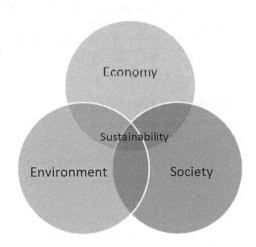

FIGURE 1.2
Triple bottom line concept of sustainable development.

the (a) identification of the impacts that result from human activities and (b) indicators, tools, and procedures needed to avoid, minimize, and/or mitigate these impacts through more sustainable design practices. These impacts will be elaborated in the following chapters.

The 27 principles in the 1992 Rio Declaration highlight the importance of protection of environmental quality while meeting the needs of population growth. These principles were reinforced in the 2002 World Summit on

Sustainable Development (WSSD) in Johannesburg. Principles 1, 3, and 4 of the Rio Declaration state that:

- "Human beings are at the centre of concerns for sustainable development. They are entitled to a healthy and productive life nature" (Principle 1).
- "The right to development must be fulfilled so as to equitably meet developmental and environmental needs of present and future generations" (Principle 3).
- "In order to achieve sustainable development, environmental protection shall constitute an integral part of the development process and cannot be considered in isolation from it" (Principle 4).

Many believe that such statements are not practical or feasible scientifically as development incurs the depletion of resources that cannot be sustainable. However, the WSSD of 2002 and the Rio Summit +20 in 2012, declarative statements of principles and others that have followed, show the need to develop the knowledge and tools to address the goals of sustainability. Although more than 190 nations agreed in the latter summit for better global environmental management and protection of the oceans and food security and promotion of a green economy, many have criticized the lack of detail and ambition.

The most recent set of goals for 2030 (https://sustainabledevelopment.un.org/rio20) aim to:

1. End poverty in all its forms everywhere
2. End hunger, achieve food security and improved nutrition, and promote sustainable agriculture
3. Ensure healthy lives and promote well-being for all at all ages
4. Ensure inclusive and equitable quality education and promote lifelong learning opportunities for all
5. Achieve gender equality and empower all women and girls
6. Ensure availability and sustainable management of water and sanitation for all
7. Ensure access to affordable, reliable, sustainable and modern energy for all
8. Promote sustained, inclusive and sustainable economic growth, full and productive employment, and decent work for all
9. Build resilient infrastructure, promote inclusive and sustainable industrialization, and foster innovation
10. Reduce inequality within and among countries
11. Make cities and human settlements inclusive, safe, resilient, and sustainable

12. Ensure sustainable consumption and production patterns
13. Take urgent action to combat climate change and its impacts
14. Conserve and sustainably use the oceans, seas, and marine resources for sustainable development
15. Protect, restore, and promote sustainable use of terrestrial ecosystems, sustainably manage forests, combat desertification, and halt and reverse land degradation and halt biodiversity loss
16. Promote peaceful and inclusive societies for sustainable development, provide access to justice for all and build effective, accountable, and inclusive institutions at all levels
17. Strengthen the means of implementation and revitalize the global partnership for sustainable development

For example, one of the most relevant for sustainable engineering is goal 3.9, which indicates that by 2030 there will be substantial reduction in the number of deaths and illnesses from hazardous chemicals and air, water and soil pollution, and contamination. Goals 6, 7, 9, 11, and 13 are most important for engineers. More detail is shown in Appendix A. Furthermore, to address climate change, countries adopted the Paris Agreement at the COP21 in Paris on December 12, 2015 (UNFCCC 2015). The Agreement entered into force shortly thereafter, on November 4, 2016. In the Agreement, all countries agreed to work to limit global temperature rise to well below 2°C and, given the grave risks, to strive for 1.5°C.
Relevant elements include:

- To achieve this temperature goal, parties aim to reach global peaking of greenhouse gas emissions to achieve a balance between anthropogenic sources and removals by sinks of GHGs.
- Developed countries should take the lead.
- Sinks and reservoirs should be conserved and enhanced.
- Enhancing adaptive measures.

In summary, these agreements pose challenges and opportunities for engineers for design of infrastructure, processes, and projects.

1.3 Challenges Faced by Engineers and Their Responsibilities

Since engineers are builders and problem solvers in industry and society, it is reasonable that engineering education is an excellent platform for imparting additional skills that can address contemporary challenges worldwide.

Multi- and interdisciplinary approaches are necessary to address these complex social, economic, and technological challenges. These approaches can effectively complement the result-oriented analytical approach to problem-solving that engineers receive as an integral part of their rigorous training. There also exists a clear trend toward multidisciplinary education in all fields of engineering: de Graaff and Ravesteijn (2001) describe the crucial need for the "complete engineer," an individual who not only has technical-scientific skills, but also an understanding of the interplay between technology and society, organizational and management skills, as well as social and communications skills.

Engineers, for example, often select materials for infrastructure or other processes. Infrastructure is highly important, as it is needed by humans to live in urban settings, shelter them from environmental risks, and protect the environment from wastes. Some materials such as minerals are not renewable and may be depleted eventually. Others such as wood are considered renewable. Other materials for construction such as concrete contain a wide variety of materials. Thus, the choice of materials can have significant impacts on resources. Past practices are not sufficient for the changing world. Engineers provide solutions to problems for the real world. However, there are many constraints for engineers as the world is becoming more and more complex. New technologies are being developed. Numerous sciences including biological and social must be considered by engineers. Engineers must work with people of various cultures and within many different regulations. Different cultures prioritize aspects differently, for example, such as access to clean water over considerations of climate change. Environmental standards may not be stringent in the developing country so higher than local standards must be applied or practices or materials modified for local conditions. The World Federation of Engineering Organizations (WFEO) has developed the Model Code of Ethics and the Model Code of Practice for Adaption to Climate Change (WFEO 2015). The purpose of these codes is to ensure that ethics based on universal values are practiced but modified to local conditions.

Engineers must be able to understand the implications of their work in social, economic, and environmental contexts. Budgets and business plans have always impacted engineering projects. More recently, energy efficiency and reduced environmental impact have been incorporated into design. Tools have been developed to assist in this process such as sustainability and life cycle assessments that will be discussed later. Indicators for social and cultural impacts are more difficult to assess quantitatively and are subject to viewpoint. For example, a mining company might see the project as beneficial to the community for employment. However, the local community might view that the project is negatively impacting their cultural values or land.

Allenby (2012) has suggested that engineers should be able to identify potential social and cultural issues of a project or process, enable communication to address concerns, reduce impacts as much as possible, and communicate the changes to the concerned groups before finalizing the work plan. Therefore, engineers now have to think beyond the traditional aspects that engineers have been used to. Many things have impacted the natural environment. Safe service and cost effectiveness have been the main drivers for engineering design without the concern for material or energy reduction. Technologies are more complicated and solutions are thus more complex. Engineers have to now think about environmental and social systems over the whole life of a project. Environmental, social, and economic concerns must all be balanced while meeting technical demands. These are the challenges and responsibilities of the present and future engineers, particularly in light of global warming. Reduction in greenhouse gas emissions is essential for slowing climate change. For example, Ainger and Fenner (2014) indicated that more natural flood defense schemes such as natural wetlands are now being considered more frequently, as they are less expensive, can increase carbon sequestration, and mitigate urban heat-islands effects.

The American Society of Civil Engineers (ASCE) has recently developed three policies to reflect the new issues of climate change adaptation and mitigation (Policy Statements 360 and 488) and sustainable development (Policy Statement 418). Policy Statement 418 had indicated "the need for social equity in the consumption of resources." In addition, engineers "must actively promote and participate in multidisciplinary teams with other professionals, such as ecologists, economists, sociologists and work with the communities served and affected to effectively address the issues and challenges of sustainable development." In their problem-solving, engineers must use the most appropriate measures to achieve sustainability for society.

Yong et al. (2014) showed that to understand degradation of the environment, knowledge of the impacts of humans on the environment is required as shown in Figure 1.3. Some of the issues can be summarized as pollution of (a) water, (b) atmosphere, and (c) land.

- Loss and degradation of soil quality due to the use of pesticides, insecticides, fertilizers, and other soil amendments
- Increased use of natural resources by mining and forestry activities and energy production
- Increased greenhouse gas and other emissions, leading to acid rain and climate change that increases severe weather occurrences, water levels, and erosion of coastal areas among other effects
- Biological magnification of pollutants by plants, aquatic organisms, and animals

In addition, Yong et al. (2014) also indicated that sustainability principles require classification of resources as renewable and nonrenewable. Renewable natural resources, however, can become nonrenewable, if they are used at rates higher than they can be replaced and hence this is not sustainable.

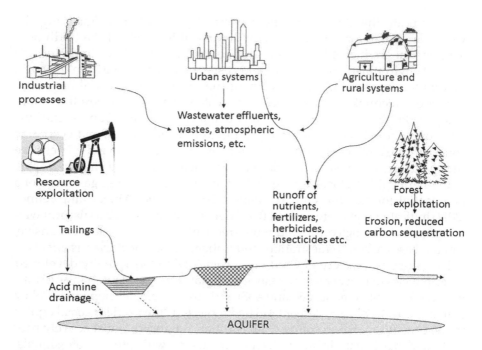

FIGURE 1.3
Environmental impacts resulting from various human activities. (Adapted from Yong et al. [2014].)

Water is a natural resource that can be easily impacted by pollution, reducing its use for human consumption.

Some other major challenges can include:

- Depletion of nonrenewable resources such as fossil fuels and minerals
- Inadequate management of industrial wastes and effluents, resulting in impacts to human health and quality of the biotic environment
- Depletion of agricultural lands occurs due to urban sprawl. High-yield practices may increase land and water pollution
- Deforestation and inadequate availability of carbon sinks contributing to increased atmospheric CO_2 and subsequent climate change

To achieve a more sustainable society, various practices (Yong et al. 2014) can be employed:

- Protection of the soil, a natural resource that is the basis of at least 90% of the production of food, energy, and materials
- Development of sustainable remediation and practices for environmental management

- Development of renewable resources to replace nonrenewable ones
- Development of protocols and procedures for technological development and new social attitudes to mitigate and adapt to climate change, in particular

1.4 Impacts on the Environment and Resource Use

Energy is used for many applications in modern society. Some applications include heating, cooling, lighting, ventilation, transportation, using electronic devices, and communications, where renewable and nonrenewable sources of energy are employed. Renewable energy sources include the use of solar, hydro, wind, and biomass, while fossil fuels (crude oil, natural gas and coal) are considered as nonrenewable due to their extensive production periods. According to the OECD/IEA (2017a), total energy consumption has doubled from 1971 to 2015 from 4244 to 9385 Mtoe. However, the breakdown according to sector has not changed significantly. In 2015, industry (37%) was the major user followed by transport (29%), residential (22%), commerce and public services (8%), agriculture/forestry (2%), and others. Most power worldwide is generated from coal (39.3%) although it has been decreasing since 2002. Nuclear power generation has been decreasing since 2000 to the current 17% whereas natural gas has been increasing since the 1990s to 22.9% in 2015. Oil on the other hand has been decreasing since the 1970s. It is currently only at about 5%. Renewables made up about 22.8% of total energy in 2015 and have been increasing since the early 2000s (particularly solar photovoltaic and wind).

Greenhouse gases are not the only emissions of energy production and consumption. Pollutants such as sulfurous oxides (SO_x), nitrous oxides (NO_x), and particulate matter are produced from the combustion of fossil fuels and biomass. These then can be converted to acid rain and ground level ozone in the atmosphere. Other impacts include flooding required by hydrodams for hydroelectricity generation that destroys habitats and alters water flows. Nuclear power has significant waste disposal issues and problems due to plant failures as in Japan. Solar panel production requires significant amounts of rare metals and waste production. Biomass fuels can lead to deforestation if wood is used as the feedstock. These impacts must be taken into consideration when choosing the appropriate energy source and minimizing its use.

The World Health Organization (WHO 2017) estimates that 8.2 million premature deaths are caused each year due to unhealthy environments, air pollution in particular. These deaths are due to stroke, heart disease, cancers, and chronic respiratory disease, now amount to nearly two-thirds of the total deaths caused by unhealthy environments. Biomass combustion is a major contributor. In addition to producing air pollution like fossil fuel combustion, most biomass has high CO_2 emissions.

Like energy, material use increases with the population growth and income. The materials can be classified into construction, industrial, metals, organics, agricultural products, and forestry products. Since World War II, use has dramatically increased. In addition, many are of limited renewability such as metals. Increased recycling and reuse is essential to address this issue. Metals can also cause harm to the environment and humans due to their toxicity.

Most of the materials for products are from the lithosphere such as stones and metals. In the United States, 95% of the materials used are nonrenewable. In developing countries, like China, the rate is increasing. Rare-earth metals, gallium and indium are needed for solar panels, neodymium for wind turbines, and dysprosium for batteries in hybrids and electric vehicles. These metals are becoming exhausted (Heinberg 2011). In addition, energy and water are required for the extraction processes.

Anthropogenic sources of contaminants include (a) various industries such as mining, manufacturing and processing, and resource exploration and exploitation and (b) human activities such as construction of buildings and infrastructure, disposal and land management of waste, agriculture. Various industries can impact the environment as follows:

1. Resource extraction and (a) the metal mining industries; (b) industries involved in extraction and processing of potash, clay minerals, phosphates; (c) the industries for extraction of aggregates, sand and rock for building material production; and (d) extraction of fossil fuels (natural gas, oil, oil sands, and coal) and uranium for the nuclear power industry.
2. Utilization of soil for agriculture and forest industries.
3. Water, groundwater, and aquifer use for hydroelectricity generation and other industries.

Some environmental impacts resulting from the various activities associated with agriculture, forestry, mining, and energy are shown in Figure 1.4. Some of the impacts will be elaborately discussed in later chapters in addition to means for pollution management, and toxicity and concentration reduction. Appropriate management practices will have to be implemented to reduce impacts as much as possible. Later chapters of this book relate to the measures that can be viewed as sustainable engineering practice.

Impacts of energy based on fossil fuels can be grouped into three categories: (a) mining, drilling for extraction, and delivery of the fossil fuel; (b) raw material conversion into energy; and (c) delivery of the energy to the consumer (e.g., pipeline, trucking, train transport). Adverse impacts can be originated from the mining and drilling operations that include mine

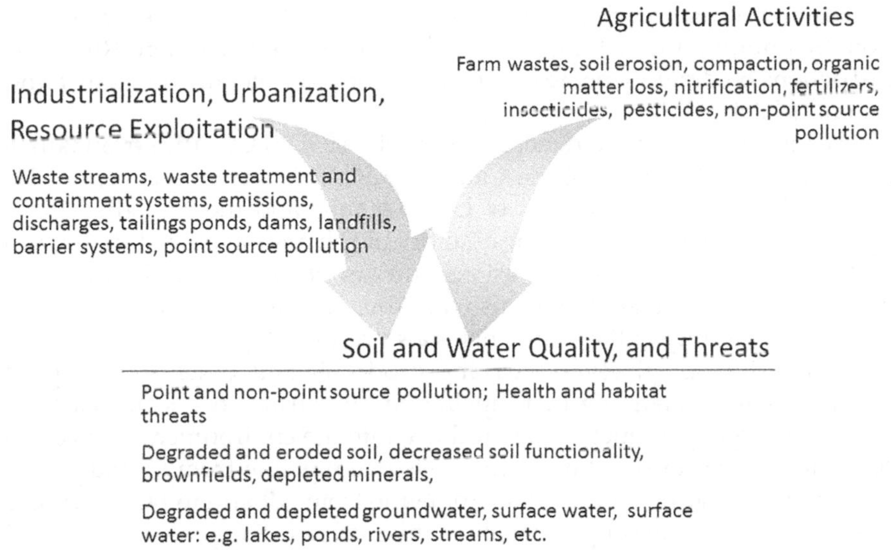

Industrialization, Urbanization, Resource Exploitation

Waste streams, waste treatment and containment systems, emissions, discharges, tailings ponds, dams, landfills, barrier systems, point source pollution

Agricultural Activities

Farm wastes, soil erosion, compaction, organic matter loss, nitrification, fertilizers, insecticides, pesticides, non-point source pollution

Soil and Water Quality, and Threats

Point and non-point source pollution; Health and habitat threats

Degraded and eroded soil, decreased soil functionality, brownfields, depleted minerals,

Degraded and depleted groundwater, surface water, surface water: e.g. lakes, ponds, rivers, streams, etc.

FIGURE 1.4
Effects of pollutant sources on soil and water quality. (Adapted from Yong et al. [2014].)

drainage, spills from mining operations, and waste material storage and streams. Spills and accidents can arise from vehicular and pipeline accidents and transmission-delivery operations. These should be avoided as much as possible through proper management of the transportation.

The Lac-Mégantic rail disaster is a key example of the hazards of rail transportation of oil (Wikipedia 2018). This event occurred in the town of Lac-Mégantic, in the Eastern Townships region of Quebec, Canada, on July 6, 2013. An unattended 74-car freight train carrying Bakken Formation crude oil rolled down a 1.2% grade hill and derailed in the downtown area. The result was a fire and explosion of multiple tank cars of the train. Forty-seven people died and more than 30 buildings, representing about half of the downtown area, were destroyed. Many also had to be demolished due to petroleum contamination. It was the fourth deadliest rail accident in Canadian history and the most fatal involving a non-passenger train.

In the first month of the disaster, firefighters and investigators worked in 15-min shifts due to heat and toxic conditions caused by hydrocarbon contamination. In the waterfront, the contamination was contained by a series of booms. Many residents were not able to return home for months as the ground around the remaining houses was contaminated with oil while those who had their homes in the most-contaminated areas were not able to return at all. Because of the extensive cleanup of the derailment area, many businesses had to relocate. The Chaudière River was contaminated by an estimated 100,000 L of oil. Some of the contamination reached the town of Saint-Georges up to 80 km. Floating barriers were installed to prevent contamination and residents were asked to limit their water consumption. Swimming and fishing were prohibited in the Chaudière River. The restrictions on drawing potable water from the river remained in effect until 2 months later.

After two phases of the remediation work (MDDELCC 2017) from 2014 to 2015 and from 2015 to 2017, the Chaudiere River is slowly recovering. The hydrocarbons in the sediments of Lac Megantic and the Chaudiere River have decreased substantially. The quantity of fish has increased but there are still high levels of deformities, lesions, tumors, etc. However, the hydrocarbons are not accumulated in the fish and thus they are safe to be eaten. The benthic community in the sediments is also recovering.

The building of pipelines carrying crude oil and other similar products across land, claiming that pipeline failures (ruptures, leaks, etc.) will discharge large quantities of the products into the environment, such as the potential future construction of the Keystone pipeline from Canada to the United States and the TransCanada pipeline from Alberta to BC in Canada, has received extensive opposition by many groups recently. Yong et al. (2014) indicated that more robust materials and engineering practices in the construction, monitoring, and operation of the pipeline are required to minimize future failures and establish a disaster response protocol for effective corrective actions in the case of a failure such as immediate remediation of the contaminated area.

Water quality and availability are major issues worldwide and thus conflicts will continue to rise. As shown in Figure 1.5, less than 5% of the global water is freshwater with only 0.2% comprising lakes and rivers, snow, ice, wetlands, and groundwater. Thus, the distribution of readily accessible freshwater is highly limited.

Some significant impacts on water quality are shown in Figure 1.6. Limitations on water quality and availability can severely impact plant and animal species that live in the water. Therefore, protection of water is highly important. Water usage by industries, for example, can produce liquid effluents that are highly toxic. These effluents must be treated before returning to the environment as shown in Figure 1.7. As indicated previously, pollutants are introduced by industrial plants, municipalities, resource extraction facilities, and agricultural activities.

Recycling or reuse of the effluents should be encouraged. Some examples of reuse include irrigation (in farming and agriculture activities), reuse in processes and energy recovery. Effluents that cannot be treated effectively and economically to reach acceptable discharge standards require storage in landfills, holding ponds, tailings ponds, or other similar systems. However, all of these containment systems have the potential to deliver pollutants to the receiving waters (ground and surface waters) because of leaks, discharges, and failures. Declining water quality decreases water availability. In the industrialized world, although water quality is in general good, water use is increasing. In the United States, it is estimated that 8 million cubic

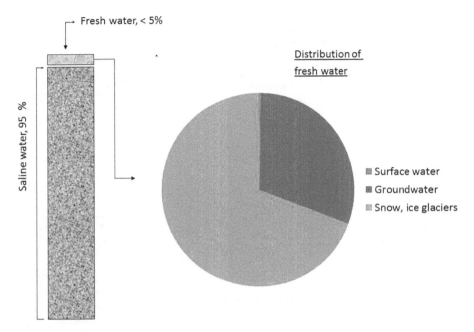

FIGURE 1.5
Distribution of fresh and saline water worldwide. (Adapted from Yong et al. [2014].)

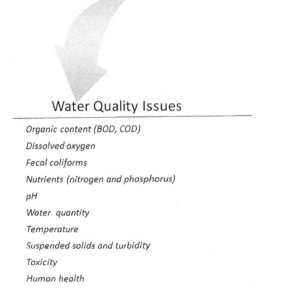

FIGURE 1.6
Issues of water quality due to contamination by pollutants.

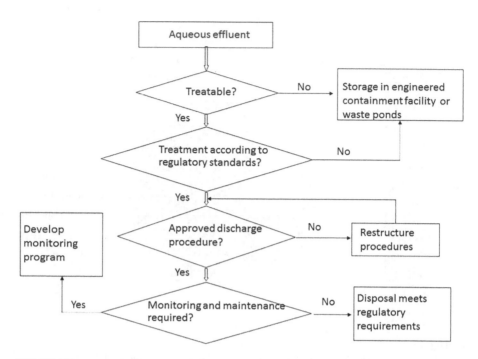

FIGURE 1.7
Management of contaminated aqueous effluents before discharge to the environment.

meters of drinking water is lost annually due to aging and leaky pipes, broken water mains, and faulty meters (Schaper 2014).

Globally water consumption has increased exponentially since the 1700s. Agriculture use is the major consumer. Water use, also, is intensive in the production of many products such as computer chips. A systems approach is needed more and more. Applicable laws and global trade policies may need to be considered from a sustainable water engineering viewpoint.

With regard to the atmosphere, ozone (O_3) is found in the stratosphere (10–50 km above the earth's surface) and is essential for absorbing ultraviolet radiation. At the earth's surface, it can also be formed from reactions of nitrogen oxides with hydrocarbons. It causes lung irritations, crop and other vegetation damage, which is a significant problem.

In the 1930s, chlorofluorocarbons (CFCs) such as CFC-13 or CFC-11 were produced for refrigeration and solvents. Molina and Roland (1974) determined that there was a link between ozone depletion and CFC use. An ozone hole over the Antarctica was found in the 1980s. Since the 1987 Montreal protocol use has been decreasing, hydrochlorofluorocarbons (HCFCs) have now replaced the CFCs in many cases. The decreased use of CFCs has markedly reduced the levels in the stratosphere and other levels.

Human activities such as combustion of fossil fuels, deforestation, industrial and agricultural activities have influenced the composition of gases in the atmosphere. In particular, carbon dioxide, water vapor, methane, nitrous oxides, CFCs, and tropospheric ozone are considered as greenhouse gases due to their absorption capabilities and long residence times in the atmosphere.

Carbon dioxide emissions have significantly increased to 33 Gt in 2015 from negligible amounts in the 1870s (OECD/IEA 2017b). In addition, its concentration has increased from 290 ppm in the preindustrial era to 403 ppm in 2016 (OECD/IEA 2017b). Regionally, China and North America account for 28% and 17%, respectively, of the global total emissions. However, as China is still developing, the emissions per capita were much lower (6.6 t CO_2) compared to the United States (15.5 t CO_2). Overall, the trends are in the opposite directions as the United States has decreased emissions by 195% from 1990 to 2015, but China has increased their emissions by a factor of threefold. Carbon dioxide accounts for 90% of the emissions compared to 9% for methane and 1% for N_2O.

Pollutants can be emitted into the atmosphere, water, and soil environment. They can accumulate in organisms or be transported or transformed in the environment. A number of mechanisms are responsible for the transport. Some of these are shown in Figure 1.8.

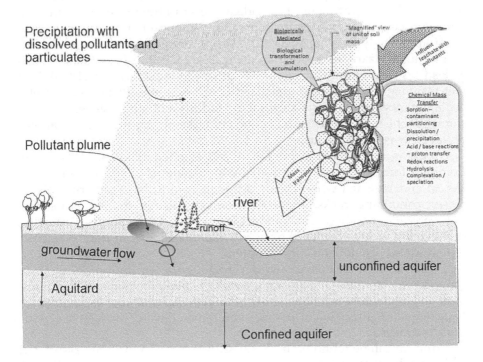

FIGURE 1.8
Transportation of pollutants to surface and groundwater. (Adapted from Yong et al. [2014].)

Air pollution can lead to significant health issues due to breathing of the pollutants. The main categories are particulates, carbon monoxide, sulfur and nitrogen oxides, and hazardous air pollutants (HAPs). Particulate matter (PM) is classified according to size. The most hazardous are those less the 10 μm (PM10) and less than 2 μm (PM2.5). The latter are particularly problematic as they can be inhaled. They can also be transported long distances in the air and they can be suspended in the air for more than a day. These particles result from construction sites, smelting of metals and incomplete combustion, incineration, or industrial and power plants. Removal of asbestos from buildings during renovations is also problematic. Carbon monoxide is emitted from incomplete combustion and is highly toxic.

Sulfur oxides are emitted from fossil fuel burning, particularly of coal (Pepper et al. 2006). Other manufacturing plants for cement, smelters, and refineries also emit sulfur dioxide. It is corrosive and can be oxidized in the atmosphere in contact with precipitation to form sulfuric acid. This leads to acid rain and can damage buildings, reduce the pH of surface water, a hazard to biota. Nitrogen oxides include nitrogen dioxide, nitrous oxide, and nitric oxide. They are commonly referred to as NO_x. They are also produced by burning of fuels, are greenhouse gases, and can lead to acid rain, like sulfur oxides.

Ozone is a priority pollutant as it is a lung irritant and can damage buildings. It is a greenhouse gas and can react in the presence of sunlight with nitrogen oxides and volatile organic compounds to form smog. The Environmental Protection Agency (EPA) has identified a total of 187 HAPs that are also hazardous. Inorganic contaminants include heavy metals such as lead (Pb), cadmium (Cd), copper (Cu), chromium, (Cr), nickel (Ni), iron (Fe), mercury (Hg), zinc (Zn), and arsenic (As). Activities that generate metal contaminants include waste disposal in landfills and others, generation and storage of chemical waste leachates and sludges, mining, and industrial processes such as metal plating.

Arsenic is a metalloid (semi-metal), but is often classified as a metal. Common forms are arsenopyrite (FeAsS), nicolite (NiAs)S, and orpiment (As_2S_3) and realgar (AsS). Arsenic originates from the weathering of the arsenic-containing rocks. Industrial sources include mining industries, dyes, preservatives and semi-conductors, and pharmaceuticals. Pesticides, herbicides, and insecticides in agriculture applications can lead to arsenic-contaminated ground and surface waters.

Arsenic is a toxic element, and a WHO regulatory limit of 50 μg/L of groundwater (aquifers) for drinking water has been adopted in many countries and regulatory agencies. In the United States, this was lowered to 10 μg/L in 2006. Ingestion of arsenic-contaminated water can lead to serious health problems.

Cadmium can be found naturally in sulfide form and is often associated with zinc, lead, or copper. Major uses of cadmium are as a filler, alloy, or active constituent for industrial products, such as nickel–cadmium batteries, enamels, fungicides, phosphate fertilizers, etc. and as a coating or plating material. Cadmium is considered to be a nonessential element from the point of view of human health effects. The EPA threshold limit for drinking water is 5 µg/L. Oral ingestion of Cd can accumulate in the liver and kidneys.

Naturally, chromium is found as ferrous chromic oxide ($FeCr_2O_4$) and lead chromate ($PbCrO_4$). It is essential for human nutrition. The chromium(III) form is found in the environment, whereas chromium compounds are generally in the chromium(VI) form which is highly toxic and considered a carcinogen. Chromium mining waste discharges and tailings ponds are other sources.

Copper is found in sandstones and other oxidized and sulfide ores. Copper can be found in airborne particle deposition from combustion of fossil fuels and wastes, industrial process effluents from production of electrical products, piping, fixtures, and different alloys, and domestic wastewater. Copper is an essential trace element for human nutrition. Threshold limits for copper vary, but are often in the range of 1.3 mg/L for drinking water and 0.1 mg/m^3 for airborne concentrations.

Lead is found in sulfide, carbonate, and oxide forms in three valence states (0, +2, and +4). The +2 form is the most common. Airborne lead deposition occurs due to emissions from waste and fuel burning. Lead can be released from smelters, coal-based power plants, battery recyclers, and incinerators. Its use was banned in the 1980s, as an additive to gasoline in the United States. Lead is considered as nonessential and can lead to severe consequences to the central nervous, kidney, and reproductive systems.

Nickel is often found in laterite deposits with sulfur oxides or sulfides. Atmospheric and waste discharges from mining activities, oil- and coal-burning power plants, and emissions from nickel-manufacturing processes are sources of nickel.

Zinc originates in mineral form as oxides, sulfides, and carbonates. Sources of airborne zinc emissions are from mining of zinc, discharges from production of zinc compounds and municipal waste treatment plant discharge, and leachates from ores.

Other contaminants of concern are salts (NaCl) from road deicing and oil drilling fluids and other industrial activities that increases water salinity. Perchlorate, a salt from explosives and rocket fuel is resistant to biodegradation and is easily transported to the environment.

Endocrine disruptors, chemicals that can replace or stimulate hormones in animals or humans through water consumption, often not removed in wastewater treatment plants. Various pesticides such as atrazine are also known as endocrine disruptors (Pepper et al. 2006).

Organic chemical contaminants in the environment can originate from industries producing various chemicals and pharmaceuticals, waste streams and disposal of chemical products, e.g., sludges and spills, and use of various chemical products as pesticides, solvents, paints, oils, etc. There are many chemicals in commercial use that end up in the environment due to human activities.

Organic chemicals found in the environment can be grouped as follows:

- Hydrocarbons including petroleum hydrocarbons, various alkanes and alkenes, and aromatic hydrocarbons such as benzene, PAHs (polycyclic aromatic hydrocarbons)
- Chlorinated hydrocarbons including TCE (trichloroethylene), carbon tetrachloride, vinyl chloride, and PCBs (polychlorinated biphenyls)
- Oxygen and nitrogen-containing organic compounds such as phenol, methanol, and TNT (trinitrotoluene)

Persistent organic pollutants (POPs) do not biologically and chemically degrade. They are also toxic, and can bioaccumulate. Some POPs include dioxins, furans, aldrin, chlordane, DDT pesticides and PAHs, and halogenated hydrocarbons. The United Nations Development Programme (UNDP 2016) listed the top 12 POPs for reduction and elimination as dioxins, furans, PCBs, hexachlorobenzene, aldrin, dieldrin, endrin, chlordane, DDT, heptachlor, mirex, toxaphene. Most are pesticides.

1.5 Concept of Sustainable Engineering

In general, conventional engineering design involves identification of a problem context, constraints, analysis and generation of alternatives, preliminary design of the chosen alternative, detailed design, startup, operation and maintenance of the process (OIQ 2003). Figure 1.9 shows the feedback loops in the various steps of conventional design. The latter steps of startup, operation and maintenance are part of the lifetime of the process but not the design phase.

To operationalize the concepts of sustainable development, engineering education and practices must adapt. This has led to the concept of sustainable engineering. Gagnon et al. (2012) defines it as "the integration of sustainability issues in the various activities related to engineering practice." Principles, guidelines, innovative designs, indicators have been put forward to form the basis of sustainable engineering practice. These will be discussed in more detail in the next chapter. Figure 1.10 shows some sustainable engineering principles (Gagnon et al. 2009). This includes the design of products, processes, buildings, infrastructure, and services. Although engineers work with many others over the lifetime of a project, engineers are solely responsible for the design.

Sustainable engineering is the process of designing or operating systems in a sustainable way that do not compromise the environment and the ability

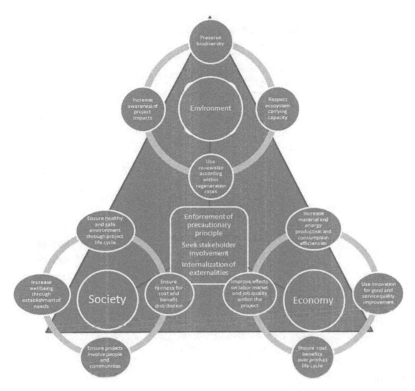

FIGURE 1.9
Steps in the conventional design process.

of the present and future generations to meet their own needs. Questions need to be asked to ensure the project is as sustainable as possible. As many engineers belong to a professional organization, they must adhere to ethics practices and sustainability. New ways of thinking are required though there are no magic formulae available to arrive at new solutions. Incorporating sustainability into engineering design has only recently been initiated due to significant challenges. Environmental, economic, and social aspects must all be integrated into the designs. This book will examine some of the concepts and tools currently available.

1.6 Conclusions

In light of the challenges posed by increasing population, constraints on resources, environmental legislation, and the environmental impact and global warming, engineers must address new ways of thinking in their designs. Sustainable engineering must address living with the planet's

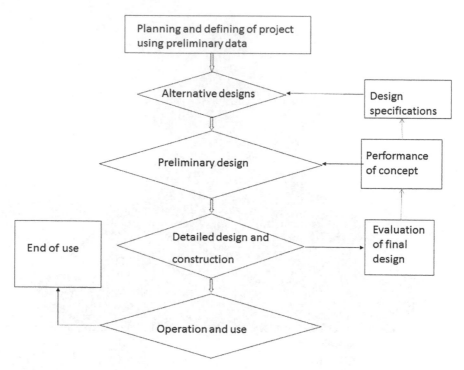

FIGURE 1.10
Sustainability principles proposed by Gagnon et al. (2009).

means, allowing all to enjoy an acceptable quality of life in the present and for future generations. With many complexities, sustainable engineering requires innovative solutions for all stages of the projects to conserve resources, reduce emissions, and impact on the environment. Complexity and uncertainty in the short and long term add new dimensions to engineering design.

References

Ainger, C. and R. Fenner, 2014. *Sustainable Infrastructure Principles into Practices*, London: ICE Publishing.

Allenby, B., 2012. *The Theory and Practice of Sustainable Engineering*, London: Pearson Education.

Butchart, S.H., M. Walpole, B. Collen, A. van Strien and J.P. Scharlemann et al., 2010. Global biodiversity: Indicators of recent declines. *Science*, 328(5982): 1164–1168.

Commoner, B., 1971. *The Closing Circle: Nature, Man and Technology*, New York: Alfred A. Knopf.

de Graaff, E. and W. Ravesteijn, 2001. Training complete engineers: Global enterprise and engineering education, *European Journal of Engineering Education*, 26: 419–427.

Elkington, J., 1998. *Cannibals with Forks: The Triple Bottom Line of 21st Century Business*, Gabriola Island, BC: New Society Publishers.

Gagnon, B., R. Leduc and L. Savard, 2009. Sustainable development in engineering: A review of principles and definition of a conceptual framework. *Environmental Engineering Science*, 26: 1459–1472.

Gagnon, B., R. Leduc and L. Savard, 2012. From a conventional to a sustainable engineering design process: Different shades of sustainability. *Journal of Engineering Design*, 23: 49–74.

Glasby, G.O., 2002. Sustainable development: The need for a new paradigm. *Environment, Development and Sustainability*, 4: 333–345.

Heinberg, R., 2011. *The End of Growth: Adapting to Our New Economic Reality*, Gabriola Island, BC: New Society Publishers.

Meadows, D.H., D.L. Meadows, J. Randers and W.W. Behrens, III, 1972. *The Limits to Growth*, New York: Universe Books.

Ministère du Développement durable, de l'Environnement et de la Lutte contre les changements climatiques (MDDELCC), 2017. Tragédie ferroviaire de Lac-Mégantic – Troisième rapport du Comité expert sur la contamination résiduelle de la rivière Chaudière par les hydrocarbures pétroliers, Québec, Direction générale du suivi de l'état de l'environnement, 7 p. et 1 ann. www.mddelcc.gouv.qc.ca/lac-megantic/rapport_chaudiere/rapport3-comite-expert-hydrocarbures.pdf. Accessed June 6, 2018.

Molina, M.J. and R.S. Rowland, 1974. Stratospheric since for chlorofluoromethanes: Chlorine atom catalyzed destruction of ozone. *Nature*, 249: 810–812.

OECD/IEA OECD/International Energy Agency, 2017a. World Energy Balances: Overview. www.iea.org/statistic.

OECD/IEA International Energy Agency, 2017b. CO_2 Emissions for Fuel Combustion. Highlights. 2017 edition OECD/IEA, Paris. www.iea.org/statistics/topics/energybalances/s/topics/energybalances/.

Ordre des ingénieurs du Québec (OIQ), 2003. *Guide de Pratique Professionnelle*, Montreal, QC: Ordre des ingénieurs du Québec.

Pepper, I.L., C.P. Gerba and M.L. Brusseau (eds), 2006. *Environmental and Pollution Science*, 2nd edition, Boston, MA: Academic Press.

Redman, C., 2014. Should Sustainability and Resilience be Combined or Remain Distinct Pursuits? www.ecologyandsociety.org/vol19/iss2/art37/. Accessed April 20, 2018.

Schaper, D., 2014. As Infrastructure Crumbles, Trillions of Gallons of Water Lost. www.npr.org/2014/10/29/359875321/as-infrastructure-crumbles-trillions-of-gallons-of-water-lost.

UNDP, 2016. Human Development Index. http://hdr.undp.org/en/content/human-development-index-hdi. Accessed July 12, 2017.

UNFCCC, 2015. Adoption of the Paris Agreement. unfccc.int/resource/docs/2015/cop21/eng/l09r01.pdf. Accessed November 11, 2018

Walker, B. and D. Salt, 2006. *Resilience Thinking: Sustaining Ecosystems and People in a Changing World*, Washington, DC: Island Press.

WFEO, 2015. World Federation of Engineering Organization, UN Scientific and Technological Communities Major Group, July 22, 2015, Version 1.6.

WHO (World Health Organization), 2017. An Estimated 12.6 Million Deaths Each Year are Attributable to Unhealthy Environments. www.who.int/mediacentre/news/releases/2016/deaths-attributable-to-unhealthy-environments/en/. Accessed April 18, 2018.

Wikipedia, 2018. Lac Mégantic Rail Disaster. https://en.wikipedia.org/wiki/Lac-M%C3%A9gantic_rail_disaster. Accessed June 6, 2018.

World Commission on Environment and Development (WCED), 1987. *Our Common Future*, Oxford: Oxford University, 400 pages.

Yong, R.N., C.N. Mulligan and M. Fukue, 2014. *Sustainable Practices in Geoenvironmental Engineering*, 2nd edition, Boca Raton, FL: CRC Press.

2

Sustainable Engineering
Theory and Concepts

2.1 Introduction

Some recent advances in engineering practice and design are related to and can form the basis of sustainable engineering practice. Engineers design many systems that require materials and energy to produce transportation, buildings, products, and other structures that can have significant impacts on environment, economy, and society. Engineers need to work with many professionals to ensure sustainable designs. Several tools and guidelines are developed. Principles and analytical methods are required to evaluate and reduce environmental, economic, and social impacts. Some of these concepts are described in this chapter. To determine if progress is being made, goals must be established for measurement of the progress, feedback, and adaptation to continually improve are all required. Indicators of some form are used to measure various components.

Various frameworks have been developed to facilitate assessment and monitoring. Some include the Global Reporting Initiative (GRI), triple bottom line, and the Natural Step. The triple bottom line includes the three environmental, social, and economic aspects as indicated previously in Chapter 1. It has been used for measuring sustainability performance for a region or business. Sustainable Seattle has used this approach (Robertson 2017).

The GRI has been used widely for measuring and reporting on the environmental, economic, and social dimensions via various indicators. It can be used for organizations of any size. Sustainability reporting guidelines include principles and performance indicators for environmental, social, and economic aspects. Although it can be very detailed, an essential element report can be chosen over a full report (GRI 2013). Reports include the organization description, management approached, indicator for each aspect, the impacts and boundaries, and management of the impact. Sustainability Tracking, Assessment and Rating System (STARS) is another reporting framework mainly used by higher education to track sustainability. It was developed by the Association for Advancement of Sustainability in Higher Education

(AASHE). The Natural Step was developed in 1989 by Dr. Karl-Henrik Robèrt (Edwards 2005) for evaluating the impact of environmental pollution on children's cancers. Various parameters to sustain life and civilization were developed through a peer review process. Four system conditions are related to (a) fossils and minerals, (b) toxins, (c) resources extraction, and (d) equity and economics (Nattras and Altomare 1999). The planning process is via graphics and metaphors with an ABCD (Awareness and Visioning, Baseline Mapping, Creative Solutions, and Decide on Priorities) approach. Universities, design firms, and corporations have used this framework.

The World Business Council for Sustainable Development (WBCSD), formerly the Business Council for Sustainable Development (WBCSD 2005) was initiated in 1992. It developed principles of eco-efficiency. They include:

- Reduction of material and energy intensity for goods and services
- Reduction of dispersion of toxic materials
- Increased ability to recycle materials
- Maximization of resource use
- Extending the durability of products
- Increasing the service intensity of goods and services

Vision 2050 was put forward to work toward sustainability by 2050. Various success measures are indicated in Table 2.1. The WBCSD, GRI, and UN Global Compact, published SDG Compass (2016) to indicate tools and indicators for businesses to work toward the Sustainable Development Goals (SDGs).

TABLE 2.1

WBCSD Vision 2050 Measures of Success

Aspect	Measure of Success
Personal values	Complete incorporation of sustainability into products, services, and life styles
Human development	End of poverty of billions of people
Economy	Internalization of carbon, ecosystem, and water costs
Agriculture	Improving water and land productivity to double agriculture output
Forests	Carbon stocks in forest doubled compared to 2010 and deforestation eliminated
Energy and power	50% reduction of carbon dioxide emissions compared to 2005 levels
Buildings	All new buildings are net zero in energy
Transportation	Near worldwide access to low-carbon and reliable mobility, infrastructure, and information
Materials	Improvement of eco-efficiency of materials and resources by 4 to 10 times compared to 2000

Source: From WBCSD (2010).

While it is for the use at the enterprise level, it can be used at various levels including the product, site, divisional or regional levels.

The World Resources Institute (WRI) and the WBCSD have been working on the development and use of the Greenhouse Gas Protocol (GFN 2018). Electronic greenhouse gas (GHG) emission calculators are found on the website (www.ghgprotocol.org), in addition to accounting and reporting standards and sector guidance. The framework is standardized to enable measurement and management of emissions from various sources.

2.2 Ecological and Other Footprints

Wackernagel and Rees (1995) developed the concept of the ecological footprint (EF) in the 1990s to indicate the impact of humans on the earth in land and sea area. It calculates the land and sea area for the resources consumed (food, wood, energy, space) and to dispose of the wastes and sequester carbon dioxide emissions (Ewing et al. 2010).

The area includes forests, fishing grounds, grazing land, built-up land, and cropland. Various assumptions must be made to estimate and simplify absorption of carbon dioxide emissions by forest. More consumption (EF) than availability (biocapacity) represents unsustainability. Although they are easy to understand, they do not take into consideration water use, pollution, economic and social aspects of sustainability among many others (Ewing et al. 2010). The current average footprint is that 1.68 earths are needed globally as of 2013 (http://data.footprintnetwork.org/#/). The Global Footprint Network website shows footprint calculators by country or region. In North America, the footprint is about 5 earths.

The impact is expressed as:

$$I = P \times A \times T \tag{2.1}$$

This equation was developed by Ehrlich and Holdren (1971). I is impact on the planet, P is population, A is affluence, and T is Technology or environmental impact per product or service used. It is a method for indicating the impact of humans on the environment. Mulder (2006) indicated that if the I factor was reduced by 50% and the population grows by 50% then T would have to be reduced by 32.4 times to improve environmental impact on the planet. As resources are consumed, the natural capacity of the earth can be exceeded. Data are also presented as the EF overshoot day. This is the day in the calendar years where the land area consumed is larger than that available. The earlier this day is, the more unsustainable the situation is.

The EF has been criticized by Fiala (2015) as improvements in agricultural productivity and the interrelations of the global economy are not taken into consideration. Methods such as land degradation and carbon dioxide emissions should be considered directly.

The energy footprint is the land area required to absorb carbon dioxide emissions and the energy required for a process or production. Although energy use is not considered, renewable energy would require less area than a nonrenewable energy. This type of footprint is the least used footprint.

The Global Footprint Network (2018) has indicated that HDI of greater than 0.7 and gha of less than 1.7 is sustainable. Only eight countries are in this range. The United States has an EF of 7 and HDI of 0.914.

Another footprint is the water footprint (WF) that takes into consideration the amount of freshwater per product per time. Water consumed in direct or indirect use or polluted is determined. It is used for surface and ground-water use, rainwater storage in soil, gray water produced, and freshwater to assimilate pollution (Galli et al. 2012). Industry has adopted this footprint more recently. It can be used to manage water supplies or production processes and in a life cycle assessment (LCA) at scales of the individual to nationwide. Businesses have used it as a sustainability indicator (Chapagain and Orr 2009).

The WF calculation includes green, blue, and gray water uses in the total WF (Hoekstra 2013). Green water originates from soil-stored water after precipitation and is thus useful for agriculture only. Blue water includes surface and groundwater and can be used for many purposes. Degradation of the water quality will disrupt future uses. LCA does not consider this. Gray water is the amount of dilution necessary to reach water quality standards. Therefore, water amounts are theoretical and higher than actual amounts. This aspect has been controversial. However, the WF has been used mainly for the food and agro-industrial sectors.

The carbon footprint examines the GHG emissions and embodied energy in terms of land required for carbon sequestration. It is becoming more common as organizations and countries try to reduce their carbon emissions to combat climate change. Often the carbon footprint is used as the amount of GHGs produced and can be used as a simplified life cycle inventory (LCI) as all life cycle stages of the product can be considered for the analysis. Large amounts of data are required as all inputs and outputs for all materials used, products and wastes must be quantified. Values are expressed as per product or service produced. Energy use can be determined from bills. However, comparisons are difficult due to the lack of standardization. The six GHGs (CO_2, CH_4, N_2O, SF_6, hydrofluorocarbons (HFCs), and chlorofluorocarbons (CFCs) are reported as direct and indirect emissions. Mainly carbon dioxide is included, and other global warming gases such as methane may not be. The reporting is in carbon dioxide equivalents. Other parameters such as biodiversity, social equity, and employment are not considered. Direct electricity, heat, steam, processing, transportation, and fugitive emissions are included. Indirect emission may originate from purchased electricity and district heating, employee transport, and services and supplier-generated emissions. International Standards Organization (ISO) 14064 (ISO 2006c) and ISO 14067 (ISO 2018) protocols are used for GHG estimates. There are numerous calculators available online such as the U.S. Environmental Protection Agency (EPA 2017). They may not lead to the desired effect of reducing carbon emissions (Nash 1988).

Material footprints are often used by companies to evaluate material consumption and waste throughout the whole production cycle. Ecological rucksack is an example that looks at the weight of a material embodied by the product arriving to the consumer via the extraction, production, and use of the material throughout its life cycle (Kibert 2012).

2.3 Sustainability Indicators

Without measurements, sustainability is only a vague concept to many that cannot be realized. Indicators are often used to measure progress. They are data that can be measured to describe a condition or trend. Some indicators include water and waste recycling rates, water and energy consumed, employment rates, GHG emissions, and water quality. General trends over time can show progress. Graphics are particularly useful for this purpose and enable communication of the results to shareholders, employees, and customers. Some management systems such as ISO 14001 require measurement. In a sustainable engineering concept, indicators help to show where and whether improvement is needed.

The types of indicators can be indicators of impacts such as footprints (ecological, water, or carbon footprints) as previously discussed. The indicators could also follow UN Sustainable Development Goals (UN 2015). For businesses, performance indicators are often used such as resource, water or energy efficiency, pollutant emissions, or waste generation. A life cycle approach over the product or process life is often employed. More information on life cycle assessments is provided in Chapter 3.

Indicator selection is important and will depend on the goals of the analyses. The indicators should provide an overall picture. Past and future trends should be covered. Data presentation is also important; for example, whether it is presented as carbon emissions or carbon emissions per metric ton of the product. Both means would be appropriate. Indicators should be measureable. Quantitative over qualitative indicators are more reliable. The indicators should be relevant to the process and not too extensive to be overwhelming. More detailed information on indicators is provided in Chapter 6.

Stahel (1982) first coined the concept of cradle-to-cradle. Two mechanisms are involved in cradle-to-cradle design: biological or technical. Biological mechanisms are designed for biodegradation, whereas technical are not. Technical mechanisms cannot return to a biological system but must remain in industry and should be durable or reusable. The third category includes heavy metals, organic pollutants that should not be used, as they are toxic. Cradle-to-cradle certification can be obtained. As for LEED certification, gold, silver, and platinum levels of cradle to cradle (C2C) can be obtained based on points obtained. The company of McDonaugh Braungart Design Chemistry (MBDC) provides the certification (Atlee and Roberts 2007).

The circular economy concept uses the cradle-to-cradle design. It is based on the "waste equals food" concept. The MacArthur Foundation (MF) has indicated that downcycling is not sufficient but a closed loop should be obtained. Many practices have encouraged disposal. However, the European Commission aims to use discarded subsidies to use technical nutrients as feed materials and promote reuse (European Commission 2011).

2.4 Industrial Ecology

An aspect that sustainable engineering has grown from is industrial ecology. The environmental and ecological approaches are emphasized. The term industrial ecology is based on the concept of industrial ecosystems by Frosch and Gallopoulous (1989). It utilizes a systems approach for environmental protection and natural resource conservation such as incorporating sustainable development objectives in processes involved in industrial production. As industry and the environment are an intertwined system, industrial ecology is comprised of (a) renewable and nonrenewable natural resource exploitation and conservation at the one end for raw materials required for industrial activity; (b) efficient industrial production through technology and resource conservation, and adherence to the 4 Rs (recycle, recovery, reduction, and reuse of waste products); and (c) environmentally conscious management of emissions and disposal of waste products from industrial activities at the other end. Industrial ecology is a holistic approach to industrial production of goods, i.e., it takes into account the goals of environment protection and resource sustainability while meeting its goals of production of goods and other life support systems to the benefit of consumers. It is also a means to mimic an ecological system (biomimicry) to design the industrial system. These goals fit into the framework of environment protection and preservation/conservation of the natural resources housed within the environment and forms the basis for the sustainable engineering practices.

Although there are a number of definitions, one of the most widely accepted is as follows: "The concept requires that an industrial system be viewed not in isolation from the surrounding systems, but in concert with them. It is a systems view in which one seeks to optimize the total materials cycle from virgin material to finished material, to component, to product to obsolete produce, and to ultimate disposal. Factors to be optimized include resources, energy and capital" (Graedel and Allenby 2003). The aim is to shift the linear model to a closed loop model to simulate nature.

According to Allenby (2012), various themes within industrial ecology are compatible with sustainable engineering. These include emphasis on:

- The system instead of specific elements
- The long-term including on regional and global scopes
- The scientific and engineering aspects related to environmental issues
- The understanding and protection of the resilience of natural and human systems
- The use of mass flow analysis (design for the environment [DFE or Dfe]) and LCA (life cycle assessment) to study energy and material flows
- Multiple scales to manage the environmental impact
- The approach of multiple scales from the material to product, to facility, to regional and global scales

Industry needed new ways to deal and increase regulatory costs, and responsibilities and requirements from multiple stakeholders. The electronics industry in particular took on this approach in Europe that included product take-back. The Netherlands broadened the approach as shown in the National Environmental Policy Plan in 1989 where a sustainable society was defined.

However, industrial ecology has mainly focused on environmental aspects as many developers involved were with an environmental background. The manufacturing sector has been the main focus with material and energy flow analyses. It also implies a comparison to biological systems with the inclusion of ecology.

Sustainable engineering must be more advanced than green engineering, which is not sufficient, although it is easier to think of physical systems other than the human aspect. Sustainable engineering changes depending on the context whether for the developed or the developing world. Engineering skills must adapt to the context. The choice of material may also influence the supply chain if a new demand is created and the product comes from a developing country.

A strategy to establish webs of industry is to set up industrial facilities in an eco-industrial park. The inputs of an industry can be the outputs of another. The industries work in symbiosis. An example of this was provided by Robertson (2017). In Kalundborg, Denmark, an eco-industrial park includes various companies working together in a 3 km radius. Heat, steam, fluid, and sludge are produced by the 1,500 MW coal-fired Asnæwas Power Station. The heat is used by a fish farm and the town's heating system, steam by Novo Nordisk and a refinery, sludge by the plasterboard plant Gyproc for its gypsum. The Statoil Refinery sends waste cooling water to the power and some flared gas to the power plant and wallboard plant. Novo Nordisk and the fish farm provide sludge to the farm fields for fertilizer (Hull et al. 2007).

There are a number of protocols and/or methodologies that have been developed to implement the objectives contained in the concept of industrial ecology. Considering industrial ecology with respect to the environment, the majority of these methodologies stem from the LCA of a particular industry or set of industries in question. The idea of using LCA or life cycle analysis of a particular item (industry, element, product, hardware, etc.) is not new or novel. LCAs are common tools for evaluation of items of interest, as, for example, economic costs over the life cycle of a particular piece of equipment or hardware, material, or mass flow over the life cycle, and risks encountered, i.e., risk assessment of a particular set of events, items, activities, economics, etc.—over the life cycle. The illustrative example shown in Figure 2.1 for consumer goods involving metal products depicts the various entities that an LCA would include.

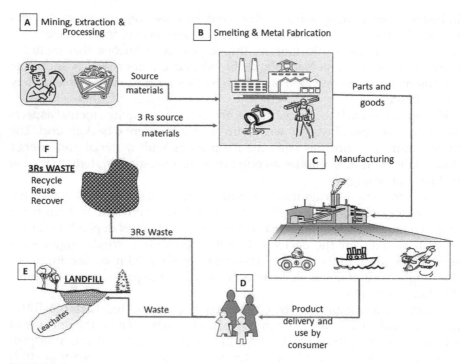

FIGURE 2.1
Illustration of life cycle events using mining and extraction of metal as an example of the start of life cycle (A), material and parts production (B), product manufacturing (C), delivery and use (D), landfilling (E), or for recycle, reuse, and recovery and remanufacturing (F) a benefit that can serve as source material for industries.

The nature and composition of the effluents and emissions (atmospheric, liquid, and solid wastes) will be a function of (a) the product produced, (b) the nature and composition of the raw materials used, (c) the processes and materials used in the production process, and (d) treatment and management of emissions, wastes, and wastewaters.

2.5 Green Chemistry and Engineering

The Hannover Principles were one of the first attempts to design systems with natural and renewable resources and eliminate wastes (McDonough and Braugart 1992). The Augsburg Materials Declaration (2002) proposed eight aspects to be considered for sustainable production that included energy-efficient processes, use of renewable energy sources, emphasis on recycling, durability, and reuse. Anastas and Zimmerman (2003) put forward the 12 Principles of Green Engineering that incorporated the principles of design of safe and benign products for recyclability and reuse, energy and mass efficiency.

The concept of green chemistry or engineering is related to the consideration of the environment for material choice or process development (Vallero and Brasier 2008), and should not be confused with sustainable engineering as only the environmental impact is evaluated. The objectives are related to waste reduction, management and selection of materials, prevention of polluting emissions, reduced toxicity, and enhanced product development. Technologically and economically viable products and processes must protect human health and the environment.

The Sandestin Green Engineering Principles were the product of a conference. They included the concept of sustainable engineering design as "to create product that meet the needs of today in an equitable fashion while maintaining healthy ecosystems and without compromising the ability of future generation to meet their resource needs."

The principles of green engineering include:

- Designers should ensure that material inputs and outputs are low toxicity as possible.
- Prevention of waste is preferable to cleanup.
- Minimization of materials and energy consumption during separation and purification processes.
- Mass, energy, space, and time efficiency should be practiced for all products, processes, and system designs.
- Output pulled instead of input pushed practices for energy and material use should be practiced for products, processes, and systems.

- When making design choices for recycling, reuse, or beneficial disposition, embedded entropy and complexity must be seen as an investment.
- The design goal should be targeted durability.
- Designing for unnecessary capability or capacity should be seen as a design flaw.
- Disassembly and value retention should be promoted by minimizing material diversity in multicomponent products.
- Available energy and material flows should be used for the design of products, processes, and systems.
- Commercial afterlife should be included in the design of products, processes, and systems.
- Renewable material and energy sources should be employed over nonrenewable.

The Sandestin Sustainable Engineering Principles consist of (Allen and Shonnard 2012):

1. Using system analysis and environmental impact assessment tools to engineer products and processes holistically.
2. Protect and improve natural ecosystems in addition to protecting health and well-being of humans.
3. Incorporate life cycle approaches into all engineering systems.
4. All material inputs and outputs must be as safe and benign as possible.
5. Reduce natural resource depletion.
6. Prevent waste as much as possible.
7. Develop and apply engineering practices that incorporate local aspirations and cultures.
8. Develop solutions to improve technologies to achieve sustainability.
9. Engage communities and stakeholders in the development of engineering solutions.

Resilience is a concept that has gained recognition recently. This is in particular due to the importance of designing infrastructure for robustness against extreme events such as floods that can have significant social, economic, and environmental impacts. Some people have used this term synonymously with sustainability. However, there are important differences as Redman (2014) has indicated. Sustainability uses indicators to evaluate the social, environmental, and economic measures for the future. Resilience focuses are improving the adaptability to weather events including social and natural capital but does not predict for the future. While both are important, a project must be resilient for sustainability but there can also be resilience without sustainability. Engineers must ethically abide by sustainability principles in practice. Businesses are now incorporating more sustainability in their practices.

2.6 Life Cycle Concept

From a life cycle point of view, the entire life of a product or process from cradle to grave must be considered and not just the production step where most emphasis had been placed previously. This includes the processing of raw materials, manufacturing of the product, transportation, distribution, recycling, and/or final disposal. All material and energy flows must be identified to recognize environmental impacts. Assessments can be fairly simple, if abbreviated or very complex then requires specialists. The LCA is a major tool in industrial ecology and thus can also be useful in sustainable engineering. In particular, the assessments can be used to identify where changes can be made to reduce environmental impact. Many methodologies have been used, initially by SETAC (the Society of Environmental Toxicology and Chemistry) and then by ISO 14040 and ISO 14044, and also by consulting organizations and academia (ISO 2006a,b). More detail on this is provided in Chapter 3.

2.7 Design for X

Various concepts of Design for X have been developed. They include design for manufacture and assembly, design for quality, design for maintainability, design to cost, design for cost, and particular design for environment. The latter is particularly related to the sustainability concept as it aims to reduce environmental impact over the life of the product or services (ISO 2002). In particular, ISO 14062 (2002) has ranked methods of reducing environmental impact in order of decreasing preference prevention, reuse, recycling, energy recovery, and disposal.

In DFE products and processes are designed in an environmentally responsible fashion. The environment and human health and safety are to be protected and natural resources sustainably used. Hazardous materials, energy, resources, and waste must all be minimized over the life cycle of the product. The terms ecodesign, design for eco-efficiency, and life cycle design are also used. The U.S. EPA's DFE program partners with industry, academia, and environmental groups. Internationally, the World Business for Sustainable Development, an association of the CEOs from over 200 companies, has supported this concept. Metrics must be chosen to measure and assess the impact of changes to the process to determine progress and to communicate with stakeholders inside and outside the company. Some of the changes may include choosing different raw materials, products, or process steps. Goals can include reduction of toxic emissions or wastes, or reuse of materials.

Design for disassembly, design for end of life, and design for recycling or reuse also are related and can assist in the design for environment. Design for disassembly aims to enable disassembly of products to facilitate recycling or reuse. Easily separated with common tools or minimal amounts of screws or types of materials are some approaches employed. The number of plastic types was reduced as the result of the mandate by European Union for automobile recycling. Design for reuse requires the product to last so that it can be resold or donated such as clothes. Other approaches are remanufacturing for heavy equipment or take-back programs for refurbishing such as for electronics.

Design for cost or design to cost addresses the economic aspects of sustainability. Design for energy efficiency aims to reduce energy for heating, cooling, lighting, and other aspects in manufacturing. Embodied energy calculations are performed over the life cycle. The International Energy Agency (IEA 2015) has indicated that recycling and reuse can decrease energy requirements twice as much as increased energy efficiency.

Design for detoxification aims to reduce the use of toxic components. An example is the removal of lead from paints, pulp, and paper processes to remove chlorine or chromated copper arsenate (CCA) replacement in treated wood. The Ford Motor Company (Epstein 2008) replaced some of the foam in their seats with a soybean-based one. It requires less energy and produces less carbon dioxide than the petroleum-based one.

CCA is a major wood preservative used in North America for many years for lumber treatment. CCA is associated with playground equipment against fungus and insects for decks, poles, and playground equipment. Children can be directly exposed by ingesting food with hands having previously in contact with CCA-contaminated wood. There might even be a risk of cancer in children (Green Building News, February 2003 issue [oikos.com/news/2003/02.html]). Leaching of Cu, Cr, and As from the wood may occur (either in service or placed in landfills and composted material). Proper incineration procedures are necessary due to hazardous risks of open burning of this type of wood. Moghaddam and Mulligan (2008) tested CCA-treated gray pine species wood for leaching using a modified toxicity characteristic leaching procedure to determine the leaching of the three metals under various conditions. The study examined the effects of pH and temperature on the leaching of the three metals from the wood, for a 5-day period: measurable amounts of chromium, copper, and arsenic were found in the leachates. Therefore, there is a risk of soil, water, and environmental contamination by chromium, copper, and arsenic from CCA-treated wood whether in use or disposed of. Disposal must be in a lined landfill to avoid contamination of the groundwater. Alternatives of less toxicity for the wood preservation are now being used including alkaline copper quaternary (ACQ) compounds, copper azole (CuAz), ammoniacal copper zinc arsenate (ACZA), copper citrate, and copper HDO (CuHDO).

The objective of design for dematerialization is to reduce the material needed for a product. Packaging and virgin materials can be reduced. Dimensions or weights can be reduced through design.

Extending the life of a product by improving durability or providing a classical design can reduce resources. This is difficult to achieve for some items, such as cell phones, which are replaced frequently due to the short market life. Fully recyclable products can reduce this problem. Upgrading some products can extend the life, reducing the need to replace all parts. Other means are to design easily repairable or refillable products. New products will be purchased if it is too difficult or expensive to repair the product. Designing products in modules can enable replacement of the required module only, in addition to reduced inventories and simplified production processes. The design for sustainability concept involves the consideration of the environment in the design of their products (UNEP 2016) in addition to the society and economy. Health and welfare, environmental protection, and conservation are practiced.

The concept of design for multiple life cycles is related to design for disassembly and remanufacturing. According to the European Commission (2015), remanufacturing is "a series of manufacturing steps acting on an end-of-life part or product in order to return it to like-new or better performance with corresponding warranty." Added value is to be achieved as much as possible. The process, however, can be quite complex. The steps (Sundin and Bras 2005) for remanufacturing include inspection, cleaning, disassembly, reassembly, repair, and testing. Reuse and remanufacturing both aim to reduce environmental impact by extending the use of the product. Reuse, however, can be simpler than remanufacturing, as extra steps would be required such as machining and other processes to make a new product in the latter case. Energy requirements can be quite low for reuse as cleaning may be the only requirement. For remanufacturing although energy requirements will be higher, it has been estimated that they may be only 15% of the initial manufacturing process (Gray and Charter 2008). The steps in this process are summarized in Figure 2.2.

Bauer et al. (2016) has summarized design for reuse and remanufacturing (Figure 2.3). Design guidelines for reuse are quite limited and those that exist are general. This is mainly because these strategies will be product- and company specific. Initial product specifications must be taken into consideration for design for reuse and design for remanufacturing.

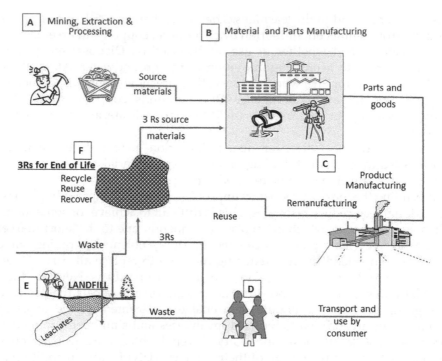

FIGURE 2.2

Modified life cycle for products. The start of life cycle is resource extraction (A), followed by material and parts production (B), the manufacturing of the product (C), delivery and use (D), and the end of the life of the product (E), where it can be then used for recycling, reuse, and recovery of parts or remanufacturing of the product (F).

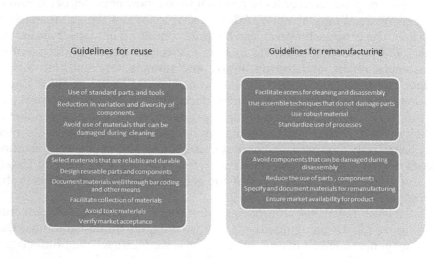

FIGURE 2.3

Guidelines for reuse and remanufacturing.

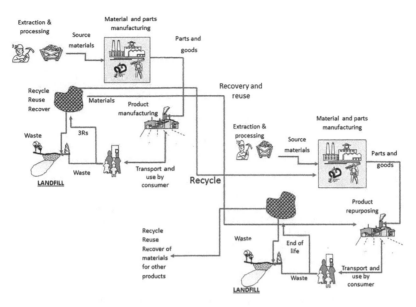

FIGURE 2.4
Repurposing of products for multiple life cycles.

Repurposing is another strategy for added value in product development. The product is sold for a different purpose. Instead of reuse and remanufacturing, new products and thus new markets that do not compete with the initial product are found. Recycling or disposal is delayed. Repurposing (Figure 2.3) can be done multiple times. A more sustainable approach to cradle to grave is the cradle-to-cradle design. This incorporates the 6 Rs as shown in Figure 2.4 (reduce, reuse, recycle, recover, redesign, and remanufacture).

Market conditions drive the profitability of product sales. A new-like product may not sell well whereas new products may. Currently, many products are designed without reuse in mind. They cannot be disassembled easily. Therefore, designing with remanufacturing is preferable. Other considerations are the collection of used products; how it is done and where the items are stored. Legal requirements are another issue.

2.8 Ecodesign

Ecodesign is the development of products with the integration of environmental performance. This followed the increased environmental awareness brought about in the 1970s and 1980s. The concept was devised in the 1990s and became more popular over the following decades. LCA was developed also in the 1990s. Material and energy efficiency were the main areas of focus.

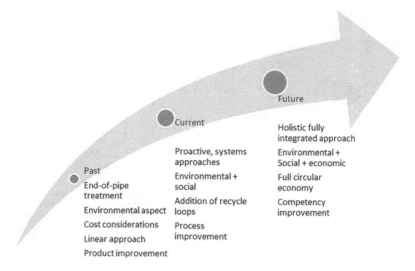

FIGURE 2.5
Evolution of ecodesign practices. (Adapted from McAloone and Pigosso 2017.)

Since 2010, there has been a shift to more high value, high-quality products, and incorporating product and service systems together as PSS. Product life can be extended through sharing. Tools and sustainability goals are being developed as businesses and organizations aim to be environmental, social, and economically oriented. More focus has been placed on design for recycling. Thus, the economy is becoming less linear and more loop oriented. Some examples are bike-sharing and car-sharing. The evolution in sustainability is shown in Figure 2.5 (McAloone and Pigosso 2017).

Recently, social innovation and the concept of circular economy in view of sustainable design of complex systems have emerged. The circular economy aims at reducing the previous make waste and emission approach by reducing material and energy consumption. Therefore, new sustainable and ecodesign approaches and business models are needed (EEA 2016). These will need to be combined with a coupled Internet of Things for digital transformation. Design for the circular economy will need various Design for X approaches including design for recycling, design for remanufacturing and design for disassembly (Achillas et al. 2013). Many challenges remain and industry will need to adopt new practices to become more sustainable.

2.9 Conclusions

Sustainable engineering practices require the development of measurable goals and strategies. For existing processes, an initial evaluation of existing conditions is performed and indicators are used to track progress over time. Various initiatives and frameworks have been developed to improve the triple bottom line. Working in teams is highly beneficial for determining the most appropriate indicators and developing sustainable engineering plans. Technological innovation is an absolute requirement to enable more sustainable practices.

Design for reuse and remanufacturing can add value to products and is in more demand. Reuse is often subjected to logistical limitations while constraints on product quality can limit remanufacturing. Reusing products in applications different from the original ones is repurposing. More work is needed for optimal design of products with the whole life cycle in mind.

References

Achillas, C., D. Aidonis and C. Vlachokostas et al., 2013. Depth of manual dismantling analysis: A cost-benefit approach. *Waste Management*, 33: 948–956.

Allen, D.T. and D.R. Shonnard, 2012. *Sustainable Engineering: Concepts, Design and Case Studies*, Upper Saddle River, NJ: Prentice Hall.

Allenby, B., 2012. *The Theory and Practice of Sustainable Engineering*, Upper Saddle River, NJ: Prentice Hall.

Anastas, P. and J.B. Zimmerman, 2003. Design through the twelve principles of green engineering. *Environmental Science and Technology*, 37(5): 94A–101A.

Atlee, J. and T. Roberts, 2007. Cradle to cradle certification: A peek inside MBDC's Black Box. *Environmental Building News*, (February 1). www.buildinggreen. com/feature/cradle-cradle-certification-peek-inside-mbdcs-black-box. Accessed November 11, 2018.

Bauer, T., G. Mandil, E. Naveux, and P. Zwolinkski, 2016. Lifespan extension for environmental benefits: A new concept of products with several distinct usage phases. *Oricedua CIRP*, 47: 430–435.

Chapagain, A.K. and S. Orr, 2009. An improved water footprint methodology linking global consumption to local water resources: A case of Spanish tomatoes, *Journal of Environmental Management*, 90: 1219–1228.

Edwards, A.R., 2005. *The Sustainability Revolution: Portrait of a Paradigm Shift*, Gabriola Island, BC: New Society Publishers.

EEA, 2016. *Circular Economy in Europe: Developing the Knowledge Base*, Luxembourg: Publication Office of the European Union.

Ehrlich, J.R. and J.P. Holdren, 1971. Impact of population growth. *Science*, 171: 1212–1217.

Environmental Protection Agency (EPA), 2017. Greenhouse Gas Equivalencies Calculator. www.epa.gov/energy/greenhouse-gas-equivalencies-calculator. Accessed November 12, 2018.

Epstein, M.J., 2008. *Making Sustainability Work: Best Practices in Managing and Measuring Corporate, Social, Environmental and Economic Impacts*, San Francisco, CA: Berret-Koehler Publishers.

European Commission, 2011. *Roadmap to a Resource Efficient Europe*, Brussels, Belgium: European Commission.

European Commission, 2015. Closing the loop—An EU action plan for the Circular Economy, COM(2015) 614 final, Brussels, December 02, 2015.

Ewing, B., D. Moore, S. Goldfinger, A. Oursler, A. Reed and M. Wackernagel, 2010. *The Ecological Footprint Atlas 2010*, Oakland, CA: Global Footprint Network.

Fiala, N., 2015. The alarming environmental costs of beef. *Scientific American*, 24(2s). www.scientificamerican.com/article/the-alarming-environmental-costs-of-beef/. Accessed November 11, 2018.

Frosch, R. and N. Gallopoulo. 1989. Strategies for manufacturing. *Scientific American*, 261: 144–152.

Galli, A., T. Wiedmann, E. Ercin, D. Knoblauch, B. Ewing and S. Giljman, 2012. Integrating ecological, carbon and water footprint into a 'footprint, family' of indicators: Definition and roled in tracking human pressure on the planet. *Ecological Indicators*, 16: 110–112

GFN, 2018. Global Footprint Network. www.footprintnetwork.org/. Accessed November 13, 2018

Global Footprint Network, 2018. WorldFootprint. www.footprintnetwork.org/. Accessed May 10, 2018.

Global Reporting Initiative, 2013. *G4 Sustainability Reporting Guidelines*, Amsterdam: Global Reporting Initiative.

Graedel, T.E. and B.R. Allenby, 2003. *Industrial Ecology and Sustainable Engineering*, Boston, MA: Prentice Hall.

Gray, C. and M. Charter, 2008. Remanufacturing and product design. *International Journal of Product Development*, 6(304): 375–392.

GRI, UN Global Compact, and WBCSD, 2016. SDG Compass: The Guide for Business Action on the SDGs. https://sdgcompass.org/wp-content/uploads/2015/12/019104_SDG_Compass_Guide_2015.pdf. Accessed May 10, 2018.

Hoekstra, A.Y., 2013. *The Water Footprint of Modern Consumer Society*, London: Routledge.

Hull, R.N., C.H. Barbu and N. Goncharova 2007. *Strategies to Enhance Environmental Security in Transition Countries*, New York: Springer.

IEA, 2015. *World Energy Outlook*, Paris: IEA.

ISO, 2002. Integrating Environmental Aspects into Product Design and Development. ISO/TR 14062:2002. www.iso.org/standard/33020.html. Accessed May 13, 2018.

ISO, 2006a. ISO/DIS 4040, *Environmental Management – Life Cycle Assessment – Principles and Framework*, Geneva, Switzerland: ISO.

ISO, 2006b. ISO/DIS 14044, *Environmental Management – Life Cycle Assessment – Requirements and Guidelines*, Geneva, Switzerland: ISO.

ISO, 2006c. ISO 14064-1:2006. Greenhouse gases – Part 1: Specification with guidance at the organization level for quantification and reporting of greenhouse gas emissions and removals, Geneva, Switzerland: ISO.

ISO, 2018. ISO 14067:2018 Greenhouse gases – Carbon footprint of products - Requirements and guidelines for quantification, Geneva, Switzerland: ISO.

Kibert, C.H., 2012. *Sustainable Construction: Green Building Design and Delivery*, 3rd edition, New York: John Wiley.

McAloone, T.C. and D.C.A. Pigosso, 2017. From ecodesign to sustainable product/service-systems: A journey through research contributions over recent decades. In: *Sustainable Manufacturing, Challengers, Solutions and Implementation Perspectives*, R. Stark, G. Seliger and J. Bonvoisin (eds), Cham, Switzerland: Springer Nature, pp. 99–110.

McDonough, M. and W. Braugart, 1992. Hannover Principles. https://econation.co.nz/hannover-principles/. Accessed November 13, 2018.

McDonough, M. and W. Braugart, 2002. *Cradle to Cradle: Remaking the Way We Make Things*, New York: North Point Press.

Moghaddam, A.H. and C.N. Mulligan, 2008. Leaching of heavy metals from chromated copper arsenate treated wood. *Waste Management*, 28: 628–637.

Mulder, K.F., 2006. *Sustainable Development for Engineers*, Sheffield: Greenleaf.

Nash, R.F., 1988. *The Rights of Nature*, Madison, WI: University of Wisconsin Press.

Nattrass, B. and M. Altomare. 1999. *The Natural Step for Business: Wealth, Ecology & the Evolutionary Corporation*, Gabriola Island, BC: New Society Publishers.

Redman, C.L., 2014. Should sustainability and resilience be combined or remain distinct. Pursuits? *Ecology and Society*, 19(2): 37. www.ecologyandsociety.org/vol19/iss2/art37/.

Robertson, M. 2017. *Sustainability Principles and Practice*, 2nd edition, Abingdon, Oxon: Routledge.

Stahel, W.R., 1982. *The Product-Life Factor*, Houston, TX: Mitchell Prize Competition on Sustainable Societies.

Sundin, E. and B. Bras, 2005. Making functional sales environmentally and economically beneficial through product remanufacturing. *Journal of Cleaner Production*, 13(9): 913–925.

UN (United Nations), 2015. Sustainable Development Goals. https://sustainable-development.un.org/?menu=1300. Accessed May 10, 2018.

UNEP (United Nations Environment Programme), 2016, UNEP 2015 Annual Report. https://wedocs.unep.org/rest/bitstreams/11114/retrieve. Accessed November 14, 2018.

Vallero, D. and C. Brasier, 2008. *Sustainable Design: The Science of Sustainability and Green Engineering*, Hoboken, NJ: John Wiley & Sons Inc.

Wackernagel, M. and W. Rees, 1995. *Our Ecological Footprint: Reducing Human Impact on the Earth*, Gabriola Island, BC: New Society Publishers.

WBCSD (World Business Council for Sustainable Development), 2005. Business Solutions for a Sustainable World. www.wbcsd.org/Overview/Our-approach/Business-solutions. Accessed May 10, 2018.

WBCSD, 2010. Vision 2050. The New Agenda for Business. www.wbcsd.org/Overview/About-us/Vision2050/Resources/Vision-2050-The-new-agenda-for-business. Accessed May 10, 2018.

3

Life Cycle Assessment for Sustainability

3.1 Introduction

Initially, engineers were concerned only with the product production step. However, that has now changed, and the whole life cycle, from the choice of the raw materials to the final use, and disposal must be considered to minimize environmental effects. Companies and governments are becoming much more aware of the environmental implications of production. Another aspect is that climate change has become the focus of many and less attention has now being focused on the entire impact on the environment.

Figure 3.1 shows the evolution of environmental management practices over the last few decades. Pollution was not considered as an important problem not that long ago. Once the effects of pollutants were identified, then dilution was deemed as the solution for air or water pollution. As this did not solve the problem, then treatment solutions (end of pipe) were developed. This resolved easily identifiable contaminants. It did not address, however, other aspects such as transportation emissions for raw materials and wastes that are produced later on. More recently, processes and products are to be designed to reduce environmental emissions and reduce the need for natural

FIGURE 3.1
Progress in environmental management over the decades. (Adapted from Curran [2015].)

resources. This is a more sustainable approach as future generations are considered in the product or process design. An improved overall picture of the environmental impact (EI) of a product or process can be obtained over its life cycle.

Products, projects, and processes have life cycles from "cradle to grave" as shown in Figure 3.2 for airplanes and cars. A production process requires extraction or production of raw materials. This includes extraction of oil, minerals, and tree harvesting. Land disturbances and transport of these materials are included in this stage. If wastes can be reused, then the term "cradle to cradle" is used.

The materials are then processed to make a product or products. This phase involves three stages: conversion of the raw materials to materials, manufacturing of the materials into products, and packaging and delivery of the products to the consumer for use. With regard to raw materials, crude oil, for example, must be refined to produce ethylene, which then can be used to produce polyethylene. The manufacturing step involves conversion of the materials to a product such as an automobile or a bottle or can. The product is then packaged for delivery to the consumer. Transportation is the major impact in this stage due to energy use and waste generation. The next stage involves use, reuse, and maintenance of the product. Wastes and emissions are produced and energy is required at all stages. The final step is disposal, reuse, or recycle of the product. Waste management can include landfilling, composting, recycling, and/or incineration. Recycling of materials will reduce resource use, emissions, and waste.

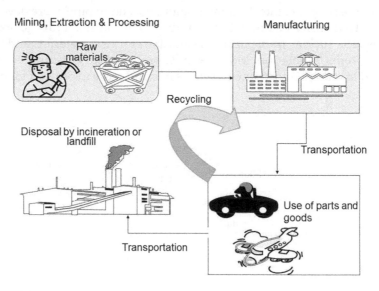

FIGURE 3.2
Life cycle of products such as automobiles and planes.

Life cycle assessment (LCA) is a tool that was developed by engineers and is used for industrial ecology to develop an understanding of EI throughout a life cycle, particularly since the 1990s. It allowed extending thinking beyond the process itself. The concept originates from nature, where raw materials form a product and then are wasted to form nutrients for other organisms. Products thus should be designed to minimize waste and reduce resource use. It is a tool that can be used for sustainable engineering to measure sustainability, evaluate more sustainable alternatives, optimize the product or process design, and indicate where improvement in the process can be made.

Data requirements can be quite extensive and there is a lack of data and tools for LCA application (Frostell 2013). Data monitoring and acquisition have not been well developed. Sometimes there is resistance to adapt to a new concept by those who are unfamiliar with the LCA methodology. However, its greatest benefit is probably for the comparison of alternatives to enable design decision-making. The Society of Environmental Toxicology and Chemistry (SETAC) and the International Standards Organization (ISO) have developed LCA methodologies. The European Commission reviewed the different LCA approaches (Reimann et al. 2012). The EC Directives such as End-of-life-Vehicles (EC 2000), Waste Electrical and Electronic Equipment (EC 2003), and Integrated Pollution Prevention and Control (IPCC) (EC 2008) incorporate LCA.

3.2 LCA Process

The LCA process consists of several steps and has now been standardized by the ISO. The ISO standard (ISO 14040 2006a) is used for the methodology of these steps. The private sector has been more receptive to this tool compared to the public sector. Engineers may not be the final decision makers but are integral in the process. The overall process is shown in Figure 3.3.

The first step is to define the boundaries of the assessment and fixing the scope. This will depend on the objective of the LCA. This could include identification where the most EI is on a process, comparing various processes to determine which has the least impact, or analysis to improve a production process on a one time or continuous basis. All relevant steps and substances need to be included. This involves defining the limits of the data collected. Wider boundaries require more data and resources, but data can be difficult to obtain. Narrower boundaries may be easier, but some important aspects can be neglected, and may not fulfill the objective of the analysis. Decisions can be made regarding whether the source of the materials should include only the supplier or further back such as mineral extraction. Particularly for commercial packages, assumptions are made and can lead to more emphasis on biodegradability, carbon footprint, or energy efficiency. Therefore, it is important to understand the assumptions, as they can bias the final results.

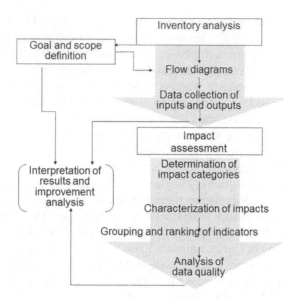

FIGURE 3.3
Process for LCA by ISO 14040 (2006a).

The basis for comparison will also need to be defined as the "functional unit." It is unique to other environmental assessment methodologies. This enables comparisons to be made between products or processes and is determined according to the goal of the assessment. If processes are to be compared, the comparison will need to be on the same basis. An example would be the amount of drinking water to be produced per year per person. The unit should not be too small; otherwise, the impact will seem small such as per day instead of per year.

The second step called "inventory analysis" is then carried out to document the input and outputs at each stage of the process. The stages include obtaining the raw materials, producing the product(s), use and maintenance of the product, and recycling/disposal of the product. Material, water, and energy inputs, emissions or outputs such as products, wastes, and by-products are tracked. A flow diagram or table can show the components included in the analysis as shown in Figure 3.4. A system may include several subsystems for a specific equipment or equipment group. Environmental burdens can be determined as shown in Azapagic (2011). In addition, the data used in the inventory must be reliable and of good quality. More and more databases are being developed to facilitate data availability, as this step can be resource intensive.

$$B_j = \sum_{i=1}^{l} bc_{j,i}x_i \qquad (3.1)$$

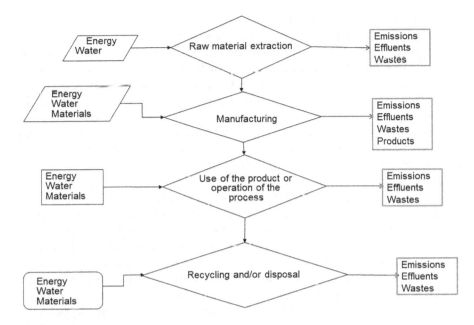

FIGURE 3.4
LCI analysis process.

Where $bc_{j,i}$ is burden j from subsystem or activity i, and x_i is mass or energy associated with that activity.

In the life cycle inventory (LCI), the data can be quite extensive in this phase. Commercial LCA tools are more available to facilitate the process. However, how the data was obtained is difficult to know. There are some public databases such as the U.S. Department of Agriculture LCA Commons available at www.lcacommons.gov (accessed February 5, 2018).

The ISO 14040 series has procedures for these, but it is general and thus can be interpreted differently by various people. All assumptions need to be documented clearly. Mass balances through a sensitivity analyses are used to validate the results. All inputs should equal the outputs. However, as some processes have only a small impact on a process, they can be neglected. Cutoff rules will express the percentage of the input–output not included in the analysis. Since processes can be complex, inputs and outputs can be partitioned in different ways. ISO standards indicate that subprocesses can be modeled or system boundaries can be expanded to avoid coproduct allocation. However, the subprocess data can be limited or system expansion is not easy. The use of physical relationships is possible such as mass, reaction enthalpy, or energy contents. If physical relationships are not applicable then economic or other considerations can be used. Guidance on allocation choice can be found in ISO (2006b). A sensitivity analysis can also be performed to determine the effect of the chosen allocation method.

Recycling can be considered in many ways. Less virgin material is required for another product once the material is recovered, processed, and used again. Coproduct allocation can be based on mass, energy, or economics of a recycled product. Boundaries can be expanded between material applications to share burdens between the produced and the recycled materials. Equal sharing is a 50:50 approach. Another approach is to neglect the environmental burdens in the initial production phase as a cutoff approach.

Expanding the boundaries to include likely aspects that will influence decisions is a consequential LCA. The example of corn ethanol was indicated by Searchinger et al. (2008). In that example, Greenhouse Gases, Regulated Emissions and Energy Use in Transportation (GREET) was used for the analysis. If the greenhouse gases (GHG) were not considered for increased corn production, then GHG emissions were 20% lower for ethanol than gasoline. However, if the additional demand for starch leads to higher prices of corn and other grains, which would lead to land use changes, then GHG emissions of ethanol production would be 47% more than gasoline.

An "impact analysis or assessment" then follows where emissions are correlated with the EI. For example, methane, carbon dioxide, and chlorofluorocarbons (CFCs) would be identified in terms of global warming potential (GWP) based on CO_2 equivalents (CO_2 eq.). Either a midpoint or endpoint can be used. Different weights are given according to the impact. These weights can be subjected to bias based on professional opinions, culture, gender, and political views. The LCI is transformed into an EI assessment. It is termed as the life cycle impact assessment (LCIA). Some impacts can include resource depletion of nonrenewables; potential for global warming; toxicity (human, terrestrial, or aquatic) and destruction of ecosystems, eutrophication, or ozone depletion; water use; potential smog production; and acidification potential. Energy use or waste production is not an impact in LCA but must be considered as GWP or other means. This can be indicated as absolute value or normalized for easier comparison. Comparisons between products, however, can be difficult in addition to resource use and toxic release impacts. For example, how can the impact on a forest in terms of resource depletion be quantified. The effect of chemicals on health is also difficult to access despite the information available. Mixtures of chemicals are particularly difficult to access. The EPA criteria for ranking of EIs (Graedel and Allenby 1995) can be according to the area impacted, the danger caused by the impact, the extent of the impact, and the risk of a wrong decision.

Various LCIA models can be used such as EPS 2000, TRACI (Tool for the Reduction and Assessment of Chemical and other Environmental Impacts), Impact World+, LIME (Life Cycle Impact Assessment Method) for impact modeling. The LCIA methods are classified into problem-oriented or damage-oriented. Problem-oriented methods incorporate EIs that are midpoint. For example, carbon dioxide emissions will lead to the category GWP. Upon aggregation, the category indicators become impact categories at the endpoint (e.g., damage to ecosystem diversity). Damage-oriented methods include impact at the endpoint such as human health impact due to the environmental burden. Eco-Indicator 99 (Goedkoop and Spriensma 2001) is a common damage-oriented method whereas CML2 is a problem-oriented method (Guinée et al. 2001). The categories can be combined into a single indicator for Eco-Indicator 99 through weighting. Impacts can be identified as potentials for global warming, ozone depletion, acidification, eutrophication, abiotic resource depletion, human toxicity, and ecotoxicity for CML2. Eco-Indicator 99 includes analysis of resources, land use, fate, exposure and effect, and damage. Damage to human health, ecosystem quality, and mineral and fossil resources are analyzed.

For the LCIA, impacts such as GWP can be calculated as the total of GHG in CO_2 equivalents per burden to give the overall impact. The impacts can be normalized, grouped, and/or weighted. All impacts can then be totaled as an overall EI. As there is a lot of discussion regarding the aggregation of the data, many indicate that the impacts can be left disaggregated for more transparency.

The last step is the "analysis or interpretation of the results of the assessment." This assessment can enable recommendations for decision-making with regard to the most environmentally friendly product if that is the objective. Alternatively, improvements in a process can be suggested if various options for production were analyzed. Pros and cons of alternatives for reducing EI can be identified. The reports must be in accordance with the objectives of the assessment. It can be difficult; however, due to the large amounts of data, assumptions are made and uncertainty in the models used. More detail can be found in ISO 14040 and 14044 (2006a,b).

The last step is the implementation of the LCA results. Sometimes it is not technically feasible to replace a material or a process. The lack of up-to-date data or lack of specificity of data is often a limiting factor. LCA should be used as a guide to show potential areas of improvement. An iterative process can be used to compare the results with previous ones to examine limitations. Additional data may be required to refine the analysis. Repetition of the analysis may be required if the goal of the analysis is not achieved.

GaBi (PE International 2008), Sima Pro (Pré Consultants 2008), CCaLC (Azapagic et al. 2010), and Ecoinvent (2007) and the U.S. LCA Commons have been developed and enable more users to perform the LCA analysis as they include background data (secondary data for materials and energy). Primary data although preferable may be difficult to obtain. This ultimately will enable improvement in the design of products, processes, and services.

LCA analyses are difficult to perform for newly developed technologies. Social, cultural, and economic dimensions are not the main objective, whereas environmental aspects are. A "stream lined LCA" may be sufficient in many cases. In this case, high-level analysis is performed with more qualitative ranking. Frostell (2013) has indicated several challenges in the use of LCA. There is no definitive link with emissions, geographical location, or time. There is no discussion on total emissions, only a few of the many material and energy flows are inputted, the various stages require various skills and substantial effort and the boundaries of the assessment can be difficult to define. These challenges can thus limit the application of this tool.

A simplified version, input-output LCA, is often used by economists. It is based on the flow of money and thus models are based on the financial linkages for the various sectors through a matrix. Energy use can be estimated as use per dollar of sales of a product. There is thus no problem of system boundaries. The main flows can be followed. The methodology by Carnegie Mello University Green Design Institute can be found out at www.eiolca.net (accessed February 5, 2018).

3.3 Life Cycle Sustainability

Sustainability analysis tools need to be more advanced than LCA, as they must integrate all components of sustainability, social and economic, in addition to the environmental of LCA. According to the United Nations Environment Programme (UNEP)/SETAC Life Cycle Initiative (2011), life cycle sustainability assessment (LCSA) is "the evaluation of all environmental, social and economic negative impacts and benefits in decision-making processes towards more sustainable products throughout their life-cycle." For an LCSA, Klöepffer (2008) indicated that:

$$LCSA = LCA + LCC + S\text{-}LCA \qquad (3.2)$$

where LCC is life cycle costing and S-LCA is social life cycle assessment.

3.3.1 LCC

ISO 15685-g was the first international standard for LCC (ISO 2008) but is restricted for buildings only. Initial and future costs can be assessed over a period of time. Hunkeler et al. (2008) divided LCC into three components: conventional, environmental, and societal. Costs with regard to the life cycle of the product are included in the conventional LCC. External costs such as environmental taxes on carbon are included in the environmental LCC. Costs associated with the impact of the production on society are included in the societal LCC, but often though these are included in the environmental LCC. Social and environmental costs can be difficult to determine.

Guidelines and code of practice for LCC are published by Swarr et al. (2011). Similar to LCA in ISO 14040, LCC usually involves four steps:

1. Definition of the goal, scope, and functional unit
2. Inventorying the costs

3. Aggregation of the cost by categories
4. Interpretation of the results

The use of LCC has been done for decades, but is often very simple (Neugebauer et al. 2016). It is often confused with other concepts such as total cost accounting (Glucha and Baumann 2004). Since there are impact analyses, it has been debated if value added can be added to indicate wealth generation or if a link via input-output modeling to gross domestic product can be included (Chang et al. 2017). These are to improve the economic sustainability analysis. Efforts will need to continue for indicators of economic impacts. Cost categories will need to be defined better. Availability of data and data quality will need to be assessed and ensured. Obtaining data along supply chains also can be quite difficult.

American Society for Testing and Materials (ASTM 2017) E917 "Standard Practice for Measuring Life-Cycle Costs of Buildings and Building Systems" is a guidance that includes:

- Development of alternative designs that are technically comparable
- Setting of the project performance periods or how long it will last
- Determination of the agency costs that includes all phases (design, construction, maintenance, rehabilitation, and demolition)
- Determination of user costs such as aesthetics, resource use
- Determination of the net present value
- Analysis of the results by sensitivity analysis and risk analysis
- Reevaluation of the design to determine the lowest cost option with the lowest impact

3.3.2 S-LCA

S-LCA examines the social and socioeconomic impact assessment of products and processes. Guidelines have been developed by UNEP/SETAC (2009). Methodological sheets were subsequently developed as a tool to design and conduct S-LCA studies and provide detailed information in the Guidelines (UNEP/SETAC 2013). A practical guide was issued on the subcategories and indicators for methodological sheets (Benoît et al. 2011). However, there is a lack of guidance on several aspects such as indicator selection, scoring methodologies, the review process, means of communication, involvement of stakeholders, and proving social impact of a product or process on a long-term basis. Various case studies of S-LCA are available such as by Franze and Ciroth (2011), but they are much fewer than for LCA.

Impact pathways of fair wages and education levels have been proposed by Neugebauer et al. (2014). S-LCA has four steps also, similar to LCA: goal and scope determination, inventory, impact assessment, and interpretation of the results.

3.3.3 LCSA

Although LCC is widely used in industry, LCSA has not been applied extensively. Indicator sets are not well developed. Neugebauer et al. (2015) have provided a three-tier approach to LCSA (Figure 3.5). The first level (Sustainability Footprint) is a low-level approach where a few selected categories are chosen such as climate change, fair wages, and production costs. It can be used as a first step for industry or other stakeholders. Tier 2 is the best practice LCSA that includes more categories for all three aspects. The full supply chain is included in the Tier 2 assessment. Tier 3 is highly advanced and included water footprints and land use. Human rights and cultural heritage and costs for environmental damage are included. The tiered approach has been practical although case studies are still limited. Hotspots can be identified in Tier 1 assessments. Trade-off may need to be considered as an environmentally beneficial technology such as recycling can pose social risks to the population doing the recycling. This can be difficult without weighting the dimensions.

Chang et al. (2017) have proposed methods to enhance the impact pathway determination for S-LCA and LCC. Due to the alignment with the sustainable development goals, fair wage and level of education should be included to evaluate social impact. Quantitative indicators are also needed.

Another approach is the new Economic Life Cycle Assessment framework (Neugebauer et al. 2016). The framework fits more with the ISO 14040 (ISO 2006a,b) requirements. An impact assessment phase is added to LCC. Two Areas of Protection (economic stability and wealth generation), five midpoint impact categories (business diversity, consumer satisfaction, long-term investment, productivity, and profitability), and two endpoint categories

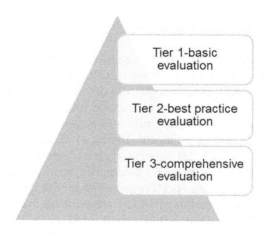

FIGURE 3.5
Tiered LSCA approach showing increasing categories as tier level increases. (Adapted from Neugebauer et al. [2015].)

(economic prosperity and resilience) are included. However, quantitative linkages for impact levels are needed.

To ensure that the social, environmental, and economic dimensions are not addressed separately in LCSA, Neugebauer et al. (2016) have proposed a conceptual framework for the Sustainability Safeguard Star. An additional three safeguard subjects are considered (economic stability, man-made environment, and social justice) (Figure 3.6). Micro and macro-economic aspects are potentially linked. Impact pathways must be further defined and more case studies are needed.

LCSA has significant potential to assist in decision-making but more research is needed. To conduct an LCSA, the steps in ISO 14040 can be followed: definition of the goal and scope, inventory analysis, impact assessment, and interpretation of results. In the first step, as LCA, LCC, and S-LCA have different purposes, a common goal and scope should be defined when these steps are to be combined. The functional unit should be related to the technical and social uses of a project, for example. System boundaries to be relevant to sustainability are shown in Figure 3.7. For the impact categories, those that are applicable across the life cycle and the three dimensions should be chosen. For example, global warming, costs of materials and disposal, and wages of workers.

For the Phase 2 inventory of LCSA, data should be collected as the unit process level in an organization. The three aspects should be consistent.

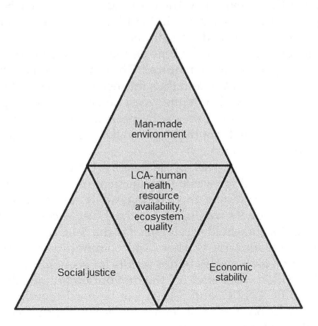

FIGURE 3.6
Relationship of LCSA to LCA. (Adapted from Neugebauer et al. [2016].)

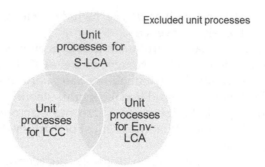

FIGURE 3.7
System boundaries for S-LCA, LCC, and environmental LCA for LSCA. (Adapted from UNEP-SETAC [2009].)

Data needs to be collected along the life cycle. S-LCA data can be much more difficult to obtain than LCA and LCC data.

For a Phase 3 assessment of impact, ISO 14040 (2006a) and ISO 14044 (2006b) should be followed. LCC does not have an impact assessment step since cost data is a direct measure. Classification in categories is possible in LCSA, but in some cases, common units may be difficult.

For Phase 4, LCSA interpretation of results can be difficult. Trade-offs between social or environmental and economic impacts may occur. An LSCD dashboard was devised by Traverso and Finkbeiner (2009) as a means of showing the results. It is a macro in MS EXCEL. Different colors show the best performance in green and worst in red.

Other LCSCA issues include the time period to be considered, stakeholder engagement, and peer review, if required. The time horizon is usually different for LCA, LCC, and S-LCA, and thus this needs to be considered. Various stakeholders such as workers/employees, consumers, local groups, and others should be consulted. A review by peers familiar with LCA and LCC is recommended, although not required in all cases. Stakeholders may need to be consulted for the S-LCA.

A case study of LCSA was presented by UNEP/SETAC (2011). Two options were evaluated for waste management of a computer. They were an informal collection system and a formal waste collection and treatment system. Eco-indicator 99 was used for the LCA, a cost-benefit analysis for the LCC, and a stakeholder category for the S-LCA for human health, wages, and job creation for workers and community groups was used. The second option gave overall better results for economic benefits, wages, and EIs. Job creation was better for the first option but health impacts are higher. Overall LCA, LCC, and S-LCA can be used together in a coordinated way in an LCSA. Each is a separate method but they can be used in an integrated manner for decision-making for environmental, social, and economic aspects.

3.4 LCA Tools

There are a number of tools that are available; but in general, they are process based and input/output analysis tools. Supply chains are used for process-based assessment. Various software packages exist to provide data for the various steps. The U.S. LCI and the GREET are a couple of examples. Others are SimaPro and GaBi. For building materials, the Inventory of Carbon and Energy (IICE) developed by Hammond and Jones (2008a,b) can be used.

Data availability and quality, site-specific variations, and definition of system boundaries are a few of the significant challenges in using LCA that can provide inconsistent results (Reap et al. 2008a,b). Social issues are not easily included in LCA. Waste material disposal in a landfill, for example, can also be problematic as it is a long-term process.

There are various international initiatives for LCA. Some of these include:

- The European Platform on Life Cycle Assessment (http:/eplca.jrc. ec.europa/eu), a collaboration between the European Commission's Joint Research Centre (JRC) Institute of Environment and Sustainability (IES) with the DG Environment, Directorate Green Economy
- The UNEP and the SETAC for the UNEP/SETAC Life Cycle Initiative (www.lifecycleinitiative.org)
- Forum for Sustainability through Life Cycle Innovation (FSLCI) (www.fslcci.org)

Ecoinvent (2007) is an international database containing thousands of datasets. The latest version 3.4 was released in October 2017. New data have been added for supply chains in the natural gas sector, chemical products, the electrical sector, and recycling of plastics.

3.5 Applications of LCA

The main applications for LCA have been in the process and manufacturing industries. It can examine the stage of the life cycle that can cause the most damage to the environment such as greenhouse gas emissions. It can also apply to services such as waste management, buildings and infrastructure, and even organizations where many products are made in different or single facility. ISO 14072 gives some guidance for organization on how to apply ISO 14040 and ISO 14044 for organizations. LCA are increasingly being implemented particularly for greenhouse gas emissions. Most assessments, however, include only emissions of CO_2, CH_4, and N_2O. Some applications of LCA are described below.

Caruso et al. (2017) examined the use of LCA for buildings. LCA can be done for each component. However, this does not take into account the links between the components. They examined residential buildings with three different structural materials for the design option. The ISO 14040 and ISO 14044 procedures were used for the environmental LCA. The steps included goal and scope definition, LCI, LCIA, and results. The life cycle of cement manufacture is shown in Figure 3.8. The reinforced concrete (RC), steel and wood structures were compared. The boundaries of the case study are shown in Figure 3.9.

SimaPro was used for collecting and analyzing the sustainability of products and services. Ecoinvent 2.2 international database was employed for the LCI and IMPACT2002+ methodology and Environmental Product Declaration (EPD) methodology for the impact assessment. The results showed that the RC option was a viable option. It had the highest impact for climate change but lower than the others for other categories throughout the entire life cycle. No option was ideal in all categories.

Simplified versions of LCA can be used in the early design stages whereas a full LCA can be performed when the design is almost complete. Various industrial sectors have employed LCA to choose the best environmental solutions for sustainable development. Some of these include the food, textile,

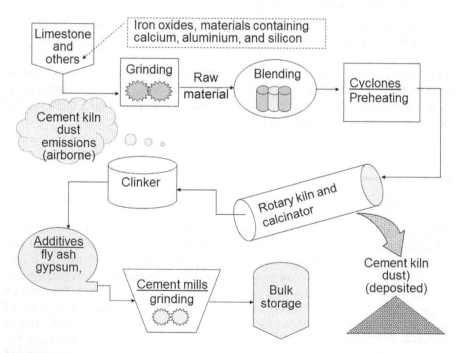

FIGURE 3.8
Basic elements in the manufacturing of cement. (Adapted from Yong et al. [2016].)

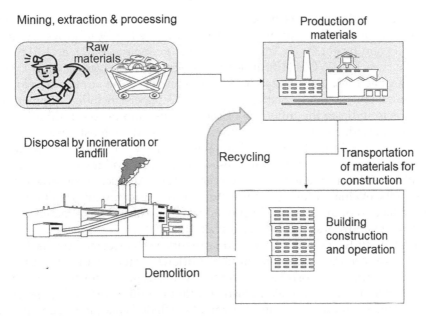

FIGURE 3.9
Boundaries of an LCA for building materials. (Adapted from Caruso et al. [2017].)

lighting, transport, paper, and auto industries (Rybaczewska-Blazejowska et al. 2016). An inverse problem solution of LCA can enable reaching the optimal solution without trial and error for designing for environment.

Applications have been provided by Azapagic (2011). For example, different electricity technologies were compared for environmental sustainability. In terms of GWP, coal was the least sustainable (1,100 g CO_2 eq./kWh) compared to nuclear resources (10 g CO_2 eq./kWh). Acidification was also significantly higher for coal (2 g SO_2 eq./kWh) compared to (0.08 g SO_2 eq./kWh) for nuclear. However, impacts like human toxicity did not vary as significantly (90 compared to 15 DCB eq./kWh). For solar energy photovoltaic and biomass, human toxicity impacts were very similar to coal. This highlights the importance of comparing various aspects not just GHG emissions over the life cycle.

Corominas et al. (2013) performed a review of 45 papers related to LCA for wastewater treatment (WWT). They indicated that there was considerable variability in the functional unit definition and the boundaries of the assessment, the methodology used and the interpretation of the results even within ISO standards. The standards need to be more strictly followed for quality and transparency purposes. The following is the evaluation of the various steps. Of the 45 studies, 23 of them included a boundary of only the operation of the plant and not the construction and demolition phases. The generation of GHG in the sludge treatment and disposal of the sludge was often not included. For the inventory phase, Ecoinvent was used. Many studies did not even include inventory data or were not detailed enough. The third step is the LCI. The studies indicated that CML was used in 19, followed by EDIP 97 (7), Eco-indicator 99 (2), Impact 2002+ (2), Ecopoints 97 (2), and EPS and ReCiPe (1 each). GWP, acidification, and eutrophication were the most used impact categories. Normalization was used in 18 of the studies by various databases. For the interpretation step, only 15 papers included a sensitivity analysis to indicate the influence of the various parameters. Limitations of the interpretations were not adequately discussed in most cases. Several new challenges in LCA for WWT are evident. They include a new focus on waste and energy reduction, water resource recovery and reuse, and nutrient recycling, the introduction of new target compounds such as organic micro-pollutants, evaluation of regional factors, and improvement of data quality and reduction of uncertainty. Improved participation of stakeholders is also required to address sustainability issues.

An LCA study of drinking water treatments was performed by Garfi et al. (2016). Five scenarios were compared: traditional water treatment to provide tap water, reverse osmosis to provide tap water, domestic reverse osmosis of traditionally treated tap water, plastic bottle mineral water, and glass bottle mineral water. Traditional water treatments included coagulation and floccu-lation, sedimentation, activated carbon adsorption, chlorine and ozone treat-ments. The functional unit of the LCA was water of $1\,m^3$. SimaPro was used with the CML 2 baseline method. Tap water by the traditional water treatment showed the least impact and was the least expensive. Electricity requirements were higher for reverse osmosis at the treatment plant compared to domestic requirements. The glass bottle mineral water had the highest impact due to raw material and energy requirements for bottle manufacturing.

Suhariyanto et al. (2017) reviewed the new concept of multiple life cycle assessment (MLCA). Products need to be designed for multiple purposes throughout its life cycle. The traditional LCA process is not capable of this type of assessment as it only considers a life cycle. There are no standards or guidelines for remanufactured products in a multiple life cycle. They found only two MLCA methodologies and both were very different in the approach. However, others used the terms such as LCA for reusable products or reus-ing, LCA for recyclable products or recycling, LCA for reuse and recycling, LCA for remanufactured products or remanufacturing. An MLCA required the extension of the life cycle for other products for transformation into a cradle-to-cradle design. The MLCA differs from an LCA with regard to the functional unit, system boundaries, and LCI. For MLCA the functional unit must be functional based instead of product based. The system boundary should be consistent with the MLCA objective. For the LCI, the most impor-tant aspects are the rates of recycling or remanufacturing.

LCA and life cycle cost assessment (LCCA) have been applied in limited cases to infrastructure projects such as road construction. The tool is named Building Environmentally and Economically Sustainable Transportation-Infrastructure-Highways (BE²ST-in-Highways) (Bloom et al. 2016).

Approximately 40% of all materials extracted in the United States are for infrastructure and buildings (Kilbert 2002). The benefits of recycling were examined in two case studies. In the Interstate-94 project, fly ash, bottom ash, foundry sand, recycled concrete aggregate (RCA), and recycle asphalt pavement (RAP) were used as recycled materials. Cost savings of $771,000 were obtained compared to the reference design. Energy use, GWP, water consumption, and hazardous waste were reduced by 37%, 39%, 255%, and 36%, respectively. In the Beltline study, RAP, recycled asphalt shingles, RCA, fly ash were used as recycled materials. The LCA/LCCA program used for this case was PaLATE. The total savings was $250,000 over the project lifetime. Energy and water use was reduced by 13% and 12%. Reductions were lower in this case than the previous case as less recycled materials were used. Therefore, the use of recycled materials can improve the sustainability of road construction. Better tracking and assessment of the use of recycled materials are needed.

3.6 Conclusions

Analytical frameworks are required to enable sustainable engineering design. LCA is one of the tools that can contribute to the analysis. It provides a more complete evaluation of the impact of product production, processes, and human activities as the whole life of a product/process is considered. The tool is becoming widely accepted and has been standardized. However, as Finkbeiner et al. (2014) have indicated there are many challenges. Weighting of the data and data analysis and new aspects such as animal welfare are challenges. Data quality and availability are problematic such as for electronic production. It can though assist in evaluating and improving sustainability. Economic and social aspects are not included in an LCA. LCC and social LCA can assist in the assessment of these. To overcome this limitation, S-LCA has been implemented which uses the triple bottom line approach to integrate LCA, LCC, and S-LCA in a coordinated approach. S-LCA is possible and should be developed more. More efforts should be made to improve the methodology and available data and more case studies are needed to facilitate the studies in order to work toward more sustainable products and processes.

References

ASTM International, 2017. Standard Practice for Measuring Life-Cycle Costs of Buildings and Building Systems, ASTM E917, 2017 edition, West Conshohocken, PA: ASTM International, 23 pages.

Azapagic, A., 2010. Life cycle assessment as a tool for sustainable management of the ecosystem services. In: *Ecosystem Services*, R.E. Hester and R.M. Harrison (eds), Issues in Environmental Science and Technology, Cambridge: Royal Society of Chemistry, vol. 30: pp. 140–168.

Azapagic, A., 2011. Assessing environmental sustainability: Life-cycle thinking and life-cycle assessment. In: *Sustainable Development in Practice: Case Studies for Engineers and Scientists*, A. Azapagic and S. Perdan (eds), Hoboken: John Wiley & Sons, pp. 56–80.

Benoît-Norris, C., G. Vickery-Niederman, S. Valdivia, J. Franze, M. Traverso, A. Ciroth and B. Mazijin, 2011. Introducing the UNEP/SETAC methodological sheets for sub-categories in social life cycle assessment (S-LCA). *International Journal of Life Cycle Assessment*, 16: 682–690.

Bloom, E.F., G.J. Horstmeier, A. Pakes Ahlman, T.B. Edil and G. Whited, 2016. *Assessing the Life Cycle Benefits of Recycled Material in Road Construction, Presented at the Geo-Chicago 2016*, Chicago, IL, vol. 269: 613–622.

Caruso, M.C., C. Menna, D. Asprone, A. Prota and G. Manfredi, 2017. Methodology for the life-cycle sustainability assessment of building structures. *ACI Structural Journal*, 114: 323–335.

Chang, Y.-J., S. Neugebauer, A. Lehmann, R. Sheumann and M. Finkbeiner, 2017. Life cycle sustainability assessment approaches for manufacturing. In: *Sustainable Manufacturing, Sustainable Production, Life Cycle Engineering and Management*, R. Shark et al. (ed), Cham, Switzerland: Springer, pp. 221–237.

Corominas, L., J. Foley, J.S. Guest, A. Hospido, H.F. Larsen, S. Morera and A. Shaw, 2013. Life cycle assessment applied to wastewater treatment: State of the art. *Water Research*, 47: 5480–5492.

Curran, M.A., 2015. Life cycle assessment: A systems approach to environmental management and sustainability. *Chemical Engineering Progress*, 111: 26–35.

Ecoinvent, 2007. Ecoinvent Database, Swiss Centre for LifeCycle Inventories. www.ecoinvent.org. Accessed February 5, 2017.

Franze, J. and A. Ciroth, 2011. A comparison of cut roses from Ecuador and The Netherlands. *International Journal for Life Cycle Assessment*, 16(4): 366–379.

Finkbeiner M., R. Ackermann, V. Bach, M. Berger, G. Brankatschk, Y.-J. Chang, M. Grinberg, A. Lehmann, J. Martínez-Blanco, N. Minkov, S. Neugebauer, R. Scheumann, L. Schneider and K. Wolf, 2014. Challenges in life cycle assessment: An overview of current gaps and research needs. In: *Background and Future Prospects in Life Cycle Assessment. LCA Compendium – The Complete World of Life Cycle Assessment*, W. Klöpffer (ed), Dordrecht: Springer, pp. 207–258.

Frostell, B., 2013. Life cycle thinking for improved resource management LCA or? In: *Handbook of Sustainable Engineering*, J. Kauffman and J.-M. Lee (eds), Dordecht: Springer Science, pp. 837–857.

Garfi, M., E. Cadena, S. Sanchez-Ramos and I. Ferrer, 2016. Life cycle assessment of drinking water: Comparing conventional water treatment, reverse osmosis and mineral water in glass and plastic bottles. *Journal of Cleaner Production*, 137: 997–1003.

Glucha, P. and H. Baumann, 2004. The life-cycle costing assessment: A conceptual discussion of its usefulness for environmental decision-making. *Building and Environment*, 39(5): 571–580.

Goedkoop, M. and R. Spriensma, 2001. The Eco-indicator 99: A damage oriented method for life cycle assessment, Methodology Report. 3rd edition, 22 June 2001, Amersfoort, The Netherlands: Pré Consultants.

Graedel, T.E. and B.R. Allenby, 1995. *Industrial Ecology*, Englewood Cliffs, NJ: Prentice Hall.

Guinée, J.B., M. Gorrée, R. Heijungs et al., 2001. *Life Cycle Assessment: An Operational Guide to the ISO Standards, Parts 1, 2a, 2b*, Dordrecht, The Netherlands: Kluwer Academic Publishers.

Hammond, G. and C.I. Jones, 2008a. Inventory of Carbon and Energy (ICE) version 1.6a University of Bath. www.organicexplorer.co.nz/site/organicexplore/files/ICE%20Version%201.6a.pdf, http://web.mit.edu/2.813/www/readings/ICE.pdf. Accessed February 5, 2018.

Hammond, G. and C.I. Jones, 2008b. Embodied energy and carbon in construction materials. *Proceedings of ICE-Energy*, 161(2): 87–98.

Hunkeler, D., K. Lichetenvort and G. Rebitzer, 2008. *Environmental Life Cycle Costing*, Boca Raton, FL: CRC Press.

ISO, 2006a. ISO/DIS 14040, *Environmental Management – Life Cycle Assessment – Principles and Framework*, Geneva, Switzerland: ISO.

ISO, 2006b. ISO/DIS 14044, *Environmental Management – Life Cycle Assessment – Requirements and Guidelines*, Geneva, Switzerland: ISO.

ISO, 2008. ISO 15686-5:2008, *Buildings and Constructed Assets-Service-Life Planning – Life Cycle Costing*, Geneva, Switzerland: ISO.

Kilbert, C.J., 2002. Policy instruments for as sustainable built environment. *Journal of Land Use and Environmental Law*, 17(2): 379–394.

Klöepffer, W., 2014. Introducing life cycle assessment and its presentation in Chapter 1. In: *LCA Compendium: The Complete World of Life-Cycle Assessment*, W. Klöepffer and M.A. Curran, Dordrecht, Germany: Springer.

Neugebauer, S., Y.-J. Chung and M. Finkbeiner, 2016. From life cycle costing to economic life cycle assessment: Introducing an economic impact pathway. *Sustainability*, 8(5): 1–23.

Neugebauer, S., J. Martinez-Blanco, R. Scheumann and M. Finkbeiner, 2015. Enhancing the practical implementation of life cycle sustainability assessment: Proposal of a tiered approach. *Journal of Cleaner Production*, 102: 165–176.

Neugebauer, S., M. Traverso, R. Scheumann, Y.-J. Chang, K. Wolf and M. Finkbeiner, 2014. Impact pathways to address social well-being and social justice in SLCA—Fair wage and level of education. *Sustainability*, 6(8): 4839–4857.

Pré International, 2008. *Gabi LCA Software and Database*, Stuttgart.

Pré Consultants, 2008. *SimaPro Database and Software*, The Netherlands.

Reap, R., F. Roman, S. Duncan and B. Bras, 2008a. A survey of unresolved problems in life cycle assessment. Part 1: Goal and scope and inventory analyses. *The International Journal of Life Cycle Assessment*, 13(4): 290–300.

Reap, R., F. Roman, S. Duncan and B. Bras, 2008b. A survey of unresolved problems in life cycle assessment. Part 2: Impact assessment and interpretation. *The International Journal of Life Cycle Assessment*, 13(5): 374–388.

Reimann, K., M. Finkbeiner, A. Horvath and Y. Matsumo, 2012. Environmental life cycle approaches for policy and decision-making support in micro and macro level applications, editors and project supervisors. In: *European Joint Commission Joint Research Center, Institute for Environment and Sustainability*, U. Pretato, D. Pennington and R. Pant (eds). http://publications.jrc.ec.europa.eu/repository/bitstream/111111111/15195/1/lbna24562enc.pdf.

Rybaczewska-Błażejowska, M., A. Masternak-Janus and W. Gierulski, 2016. Inverse problem of life cycle assessment (LCA): its application in designing for environment (DfE). *Management*, 20(2): 224–241.

Searchinger, T. R., R. Heimlich, R.A. Houghton, F. Dong, A. Elobeid, J. Fabiosa, S. Tokgoz, D. Hayes and T. H. Yu, 2008. Use of U.S. Cropland for biofuels increases greenhouse gases through emissions from land-use changes. *Science*, 219: 1238–1240.

Suhariyanto, T.T., D.A. Wahab and M.N. Ab Rahman, 2017. Multiple-life cycle assessment for sustainable products: A systematic review. *Journal of Cleaner Production*, 165: 677–696.

Swarr, T., D. Hunkeler, W. Klöpffler, H.-L. Pesonen, A. Ciroth, A.C. Brent and R. Pagan, 2011. *Environmental Life Cycle Costing: A Code of Practice*, Pensacola, FL: SETAC Press.

Traverso, M. and M. Finkbeiner, 2009. Life cycle sustainability dashboard. In: *Proceedings the 4th International Conference on Life Cycle Management*, Cape Town, South Africa.

UNEP/SETAC, 2009. *Guidelines for Social Life Cycle Assessment of Products*, Paris, France: United Nations Environment Programme.

UNEP/SETAC, 2011. *Towards a Life Cycle Sustainability Assessment*, Paris: UNEP/SETAC, 65 pages.

UNEP/SETAC, 2013. *The Methodological Sheets for Subcategories in Social Life Cycle Assessment (S-LCA)*, Paris: UNEP/SETAC.

Yong, R.N., C.N. Mulligan and M. Fukue, 2014. *Sustainable Practices in Geoenvironmental Engineering*, 2nd edition, Boca Raton, FL: CRC Press.

4

Guidelines and Frameworks Related to Sustainable Engineering

4.1 Introduction

Since sustainable development has been more widely adapted, companies have adopted new practices to achieve progress in sustainability. Guidance and standards can be regulatory or voluntary and can assist in reducing environmental and health adverse effects. At the international level, the leading nongovernmental organization for standards is the International Standards Organization or ISO. It develops various standards as well as the American Society for Testing and Materials (ASTM) International, National Institute of Standards and Technology (NIST). The EPA in the United States develops various environmental test methods and regulatory standards for water, soil, wastes, and air.

For sustainability, standards can be effective to promote sustainable practices and incorporate into policies. There are uncertainties, however, related to the standards. These should be mentioned in the standards. The standards can be for products, processes, or management systems (Sikdar et al. 2017). There are few standards on sustainability, particularly as measurability is a requirement. Therefore, significant efforts are needed. Some of the standards with respect to sustainability are described in this chapter.

4.2 ISO 14000 Standards and Environmental Management Systems

The International Organization for Standardization develops various technical and scientific standards. The technical committee (TC 176) developed the ISO 9000 series in 1987 for quality management in an organization. In 1991, the Business Council for Sustainable Development required the formulation of standards to address the impacts of business on the environment. Then TC 207 in 1993 developed the environmental series ISO 14000 series for environmental standards for impacts and responsibilities. The standards include management, environmental auditing, environmental performance evaluation, labeling, life cycle assessment (LCA), and production standards such as design for the environment. The ISO 9000 and 14000 management systems are based on the plan-do-check-act (PDCA) model of Demings and Shewhart. The reiterative cycle includes establishment of objectives and processes, implementation, measurement and evaluation followed by adaptation and improvement. The procedures must be established but no details are provided on how to do so in the standards. A company must write down the procedures, and show that they are followed closely. An auditor determines that this is done.

Various relevant standards for sustainable engineering include:

- ISO 14001: Environmental management systems – Requirements with guidance for use
- ISO 14004: Environmental management systems – Implementation general guidance on principles, systems, and support techniques
- ISO 14006: Ecodesign guidelines
- ISO 14010: Environmental auditing
- ISO 14015: Environmental assessment of sites and organizations
- ISO 14020 series: Environmental labels and declarations, general principles
- ISO 14031: Environmental performance evaluation
- ISO 14040 series: Life cycle assessment (LCA)
- ISO 14050: Terms and definitions
- SO 14060: Inclusion of environmental aspects in product standards
- ISO 14062: Integrating environmental aspects into product design and development
- ISO 14063: Environmental communication: Guidelines and examples
- ISO 14064: Measuring, quantifying, and reducing greenhouse gas emissions
- ISO 14065: Accreditation
- ISO 14067: Carbon footprint of products
- ISO 14069: Reporting of Greenhouse Gas (GHG)
- ISO 19011: Guidelines for auditing management systems (for ISO 14000 and ISO 9000)
- ISO 26000: Social responsibility

ISO 14001 is often misinterpreted (Feldman 2012). It is only eight pages long and is a process and not a performance standard. Communication to the public is not required other than the policy statement. The flowchart is developed as organization goals, environmental policy, significant aspects, objectives and targets, management programs, and finally management review.

In an environmental management system (EMS) review, aspects such as raw material and natural resource use, waste generation and disposal, and emissions and effluents are identified as significant. Methodology for identifying significance is not indicated but could include liability, frequency, severity, and public concern.

Evaluation tools such as metrics and indicators are not specified by ISO 14001 (ISO 2006). The Global Reporting Initiative (GRI), however, is more explicit. ISO 14031 has some material on this. Although this standard has not been widely adapted in the United States, Japan, and Germany, it has potential for indicator development for environmental management. Operating performance indicators, management performance indicators, and environmental condition indicators are needed for the input-output framework.

The social sustainability ISO is more guidance than strict requirements. Various stakeholders were involved in the 5-year process. They included NGOs, industry, governments, consumers, and labor organizations.

The sustainability criteria for bioenergy (ISO 13065) are currently under preparation. LCA can be used to evaluate the greenhouse gas (GHG) emissions for biofuels from biomass. Although biofuels are assumed to be carbon neutral, this may not always be the case. Agriculture and the production process require fossil fuels. In addition, if, for example, palm oil is exported to Europe from Malaysia or Indonesia for biodiesel production, it may not be carbon neutral and may even be less favorable than fossil diesel (Sikdar et al. 2017).

Brandi (2011) has proposed a conceptual process with ISO guidance that includes:

- Sustainability principles, criteria, and indicators.
- Social, economic, and environmental indicators are defined.
- The whole life cycle is to be considered for sustainability assessment.
- It is not a certification and thus there are no thresholds to reach.
- Manufacturing processes can be compared by the standard.
- Economic indicators are provided in a checklist/scorecard.

The end users can use the information to evaluate the sustainability of a product. Aspects are included as shown in Table 4.1.

Elefsiniotis and Wareham (2005) examined the link of the ISO 14000 standards to sustainable engineering of Manitoba Hydro, a provincial Crown Corporation in Canada. They identified various criticisms of the ISO 14000 series for application to Manitoba Hydro. There is confusion regarding such standards as ISO 14001, as it is a process not a performance standard.

TABLE 4.1

Sustainability Performance Aspects (Sikdar et al. 2017)

General	Social	Economic	Environmental
Measurability	Human rights	Sustainability	Water use/withdrawal
Legality	Labor rights	Fair business	GHG emissions
Monitoring/evaluation procedures	Land use rights	practices	Soil contamination
Continuous improvement	Water use rights	Financial	Air quality
Coherence/credibility	(particularly in	viability	Generation of waste
Transparency/accountability	water scarce	Market	Use of resources
Flexibility/relevance	regions)	transparency	Energy efficiency
Terminology			
Comparability			

The goal of ISO 14001 also does not guarantee optimal environmental performance but is mainly a conformance standard. In addition, companies can certify units together or separately. This can enable them to remove units that may harm their certification. In addition, companies do not actually have to comply with environmental standards to be certified under ISO 14001. Wording of the standards can be imprecise or interrupted in different ways. For example, prevention of pollution can mean end of pipe treatment instead of reduction of the source of pollution. Other words such as significant are not clearly defined. They also indicated that the term best available technology is indicated to achieve the standards but is not required, neither is the use of external reporting as a communication tool. Manitoba Hydro, however, publishes a sustainable development report.

A practical approach is needed to address the environmental concerns. For example, Manitoba Hydro came up with 13 principles for waste minimization. The approach started with elimination and reduction, then reuse and recycling followed by disposal in an appropriate manner (Manitoba Hydro 2000). Therefore, despite the vagueness of the ISO 14000 environmental series of standards, it can be an effective tool for managing environmental commitments and responsibilities as per sustainable development goals.

Material flow costing accounting (MFCA) was developed in Germany in the late 1990s. The ISO 14051 was published in 2011. The contents of this standard included the scope, terms and definitions, objectives, fundamental elements, steps for implementation, and three appendices. MFCA focuses on non-product outputs and losses of materials, in addition to products that go to market. Reducing losses and inputs reduces costs. It includes both environmental and economic aspects. According to ISO 14051 engineers are particularly required for material balance implications in various processes. ISO 14064 and ISO 14065 are also related as they manage GHG emissions.

Takakuwa (2013) has indicated that the implementation of EMS is particularly important in emerging and developing countries. Japan has successfully reduced GHG emissions, due to implementation of the environmental and energy management systems.

4.3 GRI

Elkington (1998) described the triple bottom line concept in his book *Cannicals with Forks: The Triple Bottom Line of 21st Century Business*. He indicated that businesses should be reporting environmental and social performance in addition to economic. The GRI guidance is often used for this (GRI 2016). Impacts to all three aspects must be minimized to adhere to the objectives of sustainable development as per the Brundtland Report.

The GRI is a network-based organization that has produced a worldwide sustainability reporting framework. It was setup by the United Nations Environment Programme and Coalition of Environmentally Responsible Economies (CERES). Indicators, qualitative and quantitative, are used. The GRI Sustainability Reporting Standards are to be used by organizations to show impacts on the environment, economy, and society. Stakeholders (internal and external) can then use the information for decision-making.

Table 4.2 shows the indicators used. GRI 101 is the starting point. GRI 102 and 103 report the general context of the organization and the management approach for each topic, respectively. GRI 200, 300, and 400 are the specific standards for economic, environmental, and social standards, respectively. Several principles were defined for ensuring the quality of the report: accuracy, balance, clarity, comparability, reliability, and timeliness. If the sustainability report is deemed to be according to GRI Standards, then the reporting principles must be applied, the disclosures on the organized must be given and each material topic must be identified and reported with the boundaries defined.

TABLE 4.2

Global Reporting Initiative (GRI Guidelines; www.globalreporting.org)

Indicator	Category	Aspect
Environmental	Impact	Biodiversity
		Compliance
		Effluents and waste
		Energy
		Emissions
		Materials
		Supplier environmental assessment
Economic	Direct and indirect impacts	Anti-corruption
		Anti-competitive behavior
		Economic performance
		Market presence
		Indirect economic impacts
		Procurement practices

(Continued)

TABLE 4.2 (*Continued*)

Global Reporting Initiative (GRI Guidelines; www.globalreporting.org)

Indicator	Category	Aspect
Social	Human rights	Child labor
		Disciplinary practices
		Freedom
		Forced and compulsory labor
		Human rights assessment
		Indigenous rights
		Nondiscrimination
		Strategy and management
	Labor practices	Diversity and equal opportunity
		Employment
		Freedom of association and collective bargaining
		Occupational health and safety
		Labor/management relations
		Security practices
		Training and education
		Advertising
		Customer health and safety
		Products and services
		Respect for privacy
	Product responsibility and society	Local community
		Supplier social assessment
		Public policy
		Consumer health safety
		Marketing and labeling
		Customer privacy
		Socioeconomic compliance

4.4 Other Sustainability Indices

The Dow Jones Sustainability Index (DJSI) is a measure of the sustainability performance of industries worldwide. The measurements include risk management, corporate governance, climate change mitigation, branding, supply chain, and labor standards and practices. The three measures of sustainability are included. The information is obtained from companies that pay a fee and answer questions. It is used as a public relation tool. British Petroleum (BP) was removed from the list following the DeepWater Horizon accident in 2010.

The American Institute of Chemical Engineers (AIChE) system for corporate sustainability (Chin et al. 2015) has seven indicators. The system is not used widely yet, although it uses publicly available information. The seven indicators include safety performance, social responsibility, value chain management, innovation in sustainability, product stewardship, environmental performance, and strategic commitments.

Most of the corporate sustainability indices are comparative and aim to improve sustainability and the perception of the public. Reduction of GHG emissions and use of energy, water, and nonrenewable materials are common indicators among the indices.

Standards must be reliable and repeatable to enable policy and decision-making. As there is no standard definition of sustainability, it is difficult to have an international consensus. Some quantitative indicators such as GHG emissions and energy use are common in many of the indices. Other qualitative indicators include innovation. GRI and DJSI are frameworks for sustainability of businesses.

4.5 Ecolabeling

Ecolabeling is a way to indicate to consumers that the product has some environmental or social qualities. This was on the desires of the public that products are environmentally conscious. Many claims have been made such as "ecofriendly," "recyclable," "biodegradable," and "low energy," but many are confusing and not based on credible data.

According to the ISO, the objectives are:

> through communication of verifiable and accurate information, that is not misleading, on environmental aspects of products and services, to encourage the demand for and supply of those products and services that cause less stress on the environment, thereby stimulating the potential for market-driven continuous environmental improvement.

The objectives are to protect the environment through use of renewable resources, efficient use of nonrenewable resources, reducing, reusing, and recycling of wastes and production of ecosystems and diversity. Ecolabeling programs also provide the means to improve the performance of their production processes as it provides market incentives. In addition, consumers become more conscious of environmental issues and the choices they can make to protect the environment. The ecolabel is a reliable means to assist in the evaluation of the impact of that product.

Standardization of the practices has been attempted by the ISO. There are three main types of labeling: Type I – environmental labeling, Type II – self-declaration claims, and Type III – environmental declarations. Type I is performed by a third party and provides a license for labeling products based on an LCA. Licensing fees for the use of the ecolabel symbol are paid

TABLE 4.3

Guiding Principles for Ecolabeling Programs as per ISO 14024

Voluntary
Consistent with ISO 14020 requirements
Compliance with environmental and/or relevant legislations
Considers the whole product life cycle
Consultation among interested parties in a formal process
Criteria must be selected on scientific and engineering principles to support environmental benefit claims
Transparency
Verifiable
Confidentiality maintained where required

and monitoring is performed. Type II is self-declared and may not be verifiable. Type III is a voluntary program with pre-set environmental parameters by a third party based on an LCA. A third party must do the verification. The life cycle review includes extraction of raw materials, the manufacturing process, distribution, use of the product, and then disposal.

The guiding principles for ecolabeling should be consistent with ISO 14020 and 14024. There has been growing acceptance and adoption of ISO 14000 standards by industry. The Global Ecolabelling Network (GEN) has been working on a principles document. Some of the principles are summarized in Table 4.3. Engineers are essential to the program success as they design the manufacturing processes. The main criteria that are included in the LCA are recycle content, reduction of toxicity, pollution reduction, energy efficiency, and recycling potential.

Regular review and revision of the processes is necessary to keep up with technological advances and market demands. This will lead to a continual improvement in the environmental aspect of the certified products. As the program is voluntary, the market is the main driver, not regulations. Consumers have been willing to pay more for verified products. However, the product must have other qualities and green is not the only factor in consumer choice.

Fair trade labels are Type I ecolabels. International standards must be followed for equity in pricing, labor conditions, and environmental sustainability. The Fair Trade Mark is given by the Fair Trade International Organization (also known as Fairtrade Labelling Organizations International, FLO). Fair Trade USA certifies the products in the United States with the "Fair Trade Certified Label." The International Fairtrade Mark can also be obtained.

The Ministry of the Environment, Japan (MOEJ 2009) has indicated that there are seven phases in the product life cycle (Figure 4.1):

- Resource use
- Manufacturing of the materials and parts

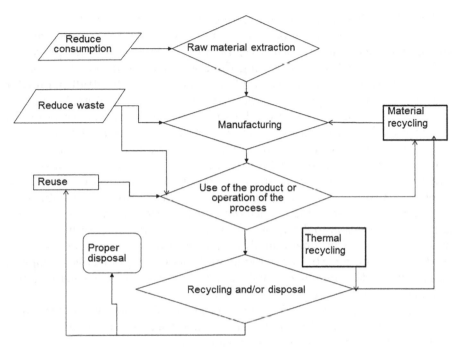

FIGURE 4.1
Japanese approach to a product life cycle. (Adapted from MOEJ [2009].)

- Manufacturing of the product
- Distribution
- Selling and purchasing
- Use
- Disposal, reuse, or recycling

LCA is used to evaluate the life cycle of a product. The procedure for LCA is in ISO 14040.

4.6 Certifications, Guidelines, and Frameworks

ISO standards have been previously discussed and thus will not be repeated. However, other certifications include some of the ISO standards. For example, Green Project Management (GPM), a member of the United Nations Global Compact (UNGC 2015b), has developed the GPM P5 standard that is free (GPM 2012). It includes ISO standards, ten principles of the UNGC (2015a), and the reporting requirements of the GRI (GPM 2012). Social, environmental, and economic aspects are included.

The ASTM E2986 is a guideline for the environmental aspects of sustainability for manufacturing. It is indicated that the sustainability must include setting the sustainability objective, identifying indicators, the processes and evaluation metrics, and setting the boundary conditions and boundaries, establishing the process, supply chains, and input-output parameter and then creating the process model. With regard to indicators, the indicators selected should be based on the sustainability objective. For example, carbon dioxide emissions for mitigation of climate change. The indicator to be used should be well defined, qualitative or quantitative, have a unit of measure, be referenced, and be specific to the level in the organization. The procedures of the evaluation of the sustainable improvement will involve data collection. Target values and a baseline should be established. To enable decision-making, indicators can be normalized, weighted, and aggregated. Decision-making can be done by simulation, optimization of various alternative, or what-if analysis. An uncertainty analysis is essential for decision-making. The reporting should include the objectives, scope of the report, the data, processes involved in the analysis, measurement results, and analysis with sensitivity/uncertainty.

The U.S. Army Corps of Engineers (USACE 2013) added sustainability to the Bid Constructability and Environmental checklist. To ensure sustainability, the following requirements must be met:

- Integrated design principles
- Optimization of energy performance
- Protection and conservation of water
- Indoor environmental quality
- Minimization of material impact through green purchasing, waste diversion, etc.
- Siting and orientation of facilities
- Layout and size of the building
- Stormwater runoff management during and after construction
- Durability and sourcing
- Transportation
- Facility performance certification for sustainability

The cradle-to-cradle certification is a promising standard for various products such as building products, food, furniture, and others. Environmental impacts and social factors are included. Environmentally preferable products are certified by a third party (Atlee and Roberts 2007). Green Seal and EcoLogo are certifiers in the United States and Canada, respectively.

The Sustainability Consortium includes many large corporations, government agencies, academics, NGOs for systems to evaluate environmental, social, and economic impacts for entire life cycle. The presence of large corporations aids in providing data that has been difficult to obtain. The Forest Sustainability Council (FSC) certifies wood products that are sustainably harvested. Third-party accredited auditors such as the Rainforest Alliance determine that chemical use, clear-cutting, forest ecosystem, and riparian management procedures are equitable. Wood products and companies can be certified with the FSC logo. Various certification programs have been developed including the Sustainable Forestry Initiative (SFI), Programme for the Endorsement of Forestry Certification (PEFC), and the Canadian Standards Association (CSA) standards.

4.6.1 Building Rating Systems

Various building rating systems (BRSs) have been developed. BREEAM was developed in the UK and Leadership in Energy and Environmental Design (LEED) in the United States and the International Green Building Tool (Bernier et al. 2010). They are used to determine the classification of the performance of the building. They establish standards for green buildings and encourage the use of environmental friendly products. Both qualitative and quantitative data are obtained for the evaluation. CEEQUAL was launched in the United Kingdom in 2003 for evaluation of the environmental and social components of infrastructure projects, and is applicable to other infrastructures than buildings. Various stakeholders including building owners, architects, and construction companies are encouraged to incorporate sustainability into the building construction.

Various tools are included for assessment of design alternatives. Others are used to evaluate performance of the building based on stars or descriptors (Ding 2008). Checklists and credit rating calculators have been developed to document and compare alternatives for design. Greater discussion and innovation for the use of new products are encouraged through the use of BRSs.

The following steps are followed in the rating systems: inputs and output classification according to impact, placing these inputs/outputs into the impact categories and then weighting these categories. The rating systems are most useful for the project design phase. Economic aspects are not included in some frameworks such as LEED, BREEAM, and LEED Canada. In addition, site-specific information may be adequately addressed such as cultural, social, and environmental variations (Ding 2008). To compensate the weighting can be adapted according to regional characteristics.

There have been various criticisms on the use of such tools. There has to be a balance between complexity and completeness. Ainger and Fenner (2014) have listed other points:

- Assessments are mainly applicable to building construction and not to other types of infrastructure projects.
- Updating must be constant to ensure assessment accuracy.

- Integrated design approaches are not addressed in the available tools.
- Various assumptions are made in the environmental impact assessments such as those that do not include behavior regarding energy conservation that can affect building performance.
- The initial building use is only considered, despite the fact that buildings can have many uses. BREEAM or LEED has an option to address this.
- Point chasing can hide various issues to be addressed.

The rating systems can, however, be used to improve sustainable engineering practices particularly in the construction industry.

The LEED rating systems include aspects related to the management of stormwater, construction waste, and air quality to achieve a sustainable building (USGBC 2018).

In the design of buildings, the U.S. Green Building Council (USGBC) LEED (USGBC 2018) certification program was developed. It is recognized worldwide and aimed to reduce environmental impacts and improved conditions for the occupants. There is a Canadian version of LEED from the Canadian Green Building Council (2018) (www.cagbc.org). It is for new and existing buildings, in many sectors such as schools, commercial, health care, warehouses and distribution centers, homes/residential, neighborhoods, and cities. Version 4 is the latest version. Impact categories are:

- Location and transport
- Sustainable sites
- Water efficiency
- Energy and atmosphere
- Materials and resources
- Indoor environmental quality

To obtain certification, compliance with all prerequisites must be documented. A minimum of 40 out of 100 points in LEED is required for the first level of certification. Higher levels of achievement are rewarded with higher levels of certification (50 points for LEED Silver, 60 points for LEED Gold, and 80 points for LEED Platinum). LEED is most useful at the design not the operation phase. Engineers need to keep developing tools such as this that can improve the economic, environmental, and social performance of buildings. Also, economic issues are not considered in the LEED. Efforts have been made to include measures in LEED to mitigate climate change (USGBC 2018). They include reduction of energy use in buildings, transportation requirements for all involved, embodied energy in water and material use, and taking into consideration land use changes that can influence carbon sequestration.

Various building ratings systems were compared by Ainger and Fenner (2014). LEED and BREEAM of the United Kingdom were introduced in the 1990s followed by the Green Building Tool (Bernier et al. 2010). The systems are to assist in decision-making with regard to increasing the sustainability of the building design. Tools are often based on LCA databases. The tools can use indicators for rating the design alternatives or performance based on descriptors.

4.6.2 Infrastructure Rating Systems

Envision (ISI 2015) is a rating system to evaluate the sustainability of infrastructure projects. Users of the system include designers, constructors, community groups, owners, and policy makers. Silver, gold, and platinum levels can be achieved based on points. The American Public Works Association (APWA), American Society for Civil Engineers (ASCE), and the American Council of Engineering Companies (ACECs) founded ISI. It is used for planning infrastructure projects related to airports, bridges, dams, roads, landfills, and water treatment systems among others. A checklist can be downloaded for a quick assessment, usually in the early stage of a project (ISI 2018a).

The Envision Rating System includes sustainability through the following:

- Climate
- Communications
- Integration with existing infrastructure
- Natural hazards
- Risk
- Safety
- Security
- Stakeholder engagement
- Synergies

Credits are given for the following:

- Energy
- Floodplain
- Invasive species
- Landscape
- Noise and light
- Pest management
- Procurement

- Stormwater
- Waste
- Water
- Team

Full implementation is the highest level in the sustainability management system. It includes:

- Ability to PDCA business processes
- Different mechanisms to manage change and project complexities
- Addresses design variable changes
- Addresses sustainability at all management levels in a project
- Includes a high degree of clarity on sustainability aspects

Sixty credits can be obtained from the categories of quality of life, leadership, resource allocation, natural world, and climate and risk (Chisholm et al. 2017). For the quality of life categories, project sustainability can be improved by enhancing health and safety, minimizing light pollution, and enhancing sustainable transportation. In the leadership category, a sustainability management system can be developed and public input can be sought. In the resource category, reduction of net embodied energy, use of local material, and waste diversion from landfill are some examples.

An Envision Sustainability Professional (ENV SP) (ISI 2018b) is trained through web-studies and passing an exam and paying a fee. This person can assist in the application of the credit application, improve the sustainability of the project, or submit a project for review. Verification of a project requires an ENV SP as part of the team.

CEEQUAL (2016) was launched in the United Kingdom in 2003. CEEQUAL and Envision are used for environmental and social performance evaluations for infrastructure projects including landscaping. CEEQUAL is available for the United Kingdom, Ireland, and international projects. It is a rating system that provides award certificates. A score is produced from answers to questions. A high degree of social and environmental performance is encouraged for civil engineering projects. CEEQUAL can be adapted for United Kingdom, Ireland, international and term projects. Certificates are provided to all partners after ratification. Ghumra et al. (2009) indicated that it does not use a life cycle approach or examine the materials used.

The systems involve input of the resources and materials used, assessment of the impact of the inputs and outputs, and weighting each category. Higher certifications can lead to high market values. However, this usually involves more resources to achieve these credits. It has also been called points hunting.

The rating systems are most applicable prior to the design phase. However, it would be more appropriate to apply before the design stage when sustainability can be really incorporated in the design. In addition, environmentally friendly materials may be very expensive to obtain. The effect of location due to local issues of water scarcity or climate, social and culture aspects is also not incorporated. This can have significant environmental issues.

Ainger and Fenner (2014) have listed several issues regarding the systems. Some include:

- Systems only minimize unsustainability and do not promote sustainability.
- Environmental aspects are mainly considered with little attention to economic and social issues.
- Few with the exception to CEEQUAL and Envision are applicable to infrastructure other than buildings.
- To maintain accuracy, regular updates are needed.
- An integrated approach is needed to enable integrated design.
- Behavioral aspects are not considered that can affect energy efficiency and other aspects.
- Most rating systems do not consider subsequent lives of the building, only the first. LEED does have "in Use" options.
- Weighting and combining unrelated criteria into a single aspect can simplify and cause difficulties in interpretation of the info. A comparison of the rating systems can be seen in Tables 4.4 and 4.5.

The Envision sustainable infrastructure rating system is a comprehensive framework of 60 sustainability criteria that address the full range of environmental, social, and economic impacts to sustainability in project design, construction, and operation. These criteria—called "credits"—are arranged in five categories: quality of life, leadership, resource allocation, natural world, and climate and risk. The full Envision guidance manual detailing the credits is provided at no cost to users. To determine the right project, Envision can be used in the earliest planning phases to evaluate infrastructure sustainability options that can result in significantly better outcomes. To do the project right, during the design and construction phases Envision provides a detailed, comprehensive set of criteria that help ensure that all significant areas of impact, as well as stakeholder views, are considered. Last but not least, when the project is complete, Envision serves as a basis for a project sustainability evaluation, helping stakeholders understand exactly how the project succeeded.

Most of the sustainable engineering issues are considered with the exception of water and energy use on affordability forced relocation and protection against corruption.

TABLE 4.4

Comparison of Rating Systems

	BREEAM	LEED	CEEQUAL	Envision
Applicability	New, renovated, operation, and maintenance of buildings (operates with CEEQUAL)	New, renovated, operation, and maintenance of buildings	Civil infrastructure, landscaping, and public realm	Civil infrastructure excluding buildings, earliest planning stages throughout operations
Technical aspect included	Site potential, energy use, water use environmentally preferred products, indoor quality, operation and maintenance practices	Site potential, energy use, water use, indoor quality, operation and maintenance practices	Various indicators for project sustainability for specific issues with targets	Five categories: quality of life, leadership, resource allocation, natural world, and climate and risk
				Various indicators for project sustainability for specific issues with targets
Verification	Detailed assessment by trained and licensed by BRE	Audit by trained assessor	Externally verified by a CEEQUAL-appointed verifier, and ratified by the Scheme Management Team	Audit by ENV SP
Ratings	Pass, good, very good, excellent, outstanding	Certified, silver, gold, platinum	Rating and certification	Envision Bronze, Silver, Gold, or Platinum
	Certificate	Award letter, certificate, and plaque	Pass – Good – Very Good – Excellent scale	
Applicability	United Kingdom	United States, Canada	United Kingdom, Ireland, Hong Kong, and international projects	United States

Questions to ask when deciding which tool to select include:

- Are all issues of sustainable engineering covered?
- Are state of the art processes supported?

TABLE 4.5

Sustainability Indicators Used in Rating or Reporting Tools

Aspects	BREEAM	CEEQUAL	LEED	GRI	Envision
Environmental					
Atmosphere	X	X	X	X	X
Biodiversity/ecology	X	X	X	X	X
Climate change	X	X	X	X	X
Energy	X	X	X	X	X
GHG management	X	X	X	X	X
Land management	X	X	X	X	X
Minimization of waste	X	X	X	X	X
Noise/dust	X	X	X	X	X
Resource/material efficiency	X	X	X		X
Soil		X			X
Water		X			X
Economic					
Bureaucracy	X		X	X	
Cost/cost-effectiveness				X	
Climate change				X	
Growth/competitiveness					
Indirect cost/impact					
Innovation investment					
Social					
Accessibility	X	X	X	X	X
Culture/communities	X	X	X	X	X
Equity	X	X		X	X
Health and safety/security		X		X	X
Heritage		X		X	X
Human rights		X			
Landscape/visual impact		X			
Participation/inclusiveness					
Stakeholder satisfaction					

- Are goals for each issue set?
- Is performance measured against the goals?
- Is the weighting system adjustable for different situations or regions?
- Does the system guarantee consistency and evaluation accuracy according to professional practice?
- Is it easily migratable to new versions?
- Is there a development process?

CEEQUAL and Envision are quite similar. CEEQUAL is not as aggressive as Envision and the points distribution depends on who is doing the evaluation (client, designer, or contractor). It is focused on issues for the United Kingdom. The sections are divided into project strategy, project management, people and communities, use of land, historic environment, ecology and biodiversity, water environment and physical resources, and transport. The questions for sustainable engineering are covered. However, issues related to local law and custom are not dealt with. A difficulty with CEEQUAL and Envision is that points are obtained based on a sustainable outcome not necessarily integrating sustainable operations into the design.

The Infrastructure Sustainability Council of Australia has developed an Infrastructure Sustainability (IS) rating tool released in 2012 (ISCA 2018). It is now called ISCA IS v1.2. Ratings are good, excellent, and leading. It covers design, construction and operation for designers, contractors, legislators, and owners. Various tools are available including scorecards for Design & As Built, and Operations and IS Materials Calculator and Guidelines.

Six themes (management and governance, use of resources, emission pollution and waste, ecology, people and place, and innovation) are covered with 15 categories to give a total of 51 issues. The questions are handled quite well. However, some issues such as human rights are not handled by the system. The weighting system is adapted to Australian conditions where water is of high importance.

4.7 Conclusions

A number of approaches exist to assist in the movement toward sustainability. The planning process involves planning, developing a vision, setting goals, and developing the methodologies to implement the vision. To aid in the implementation process, various frameworks have been developed to provide a roadmap to sustainability and to assist in the decision-making process. The selection of the framework must be appropriate for the application. For example, the GRI is most often used by organizations. Commonly used frameworks can provide more credibility since the methodology is well known. Continuous improvements, however, are being made to the frameworks to enhance their applicability. Communication is an important aspect of the process such as for the GRI for organizations to demonstrate their commitment to sustainability.

References

Ainger, C. and R. Fenner, 2014. *Sustainable Infrastructure Principles into Practices,* London: ICE Publishing.

ASTM International, 2015. Standard Guide for Evaluation of Environmental Aspects of Sustainability of Manufacturing Processes, E2986-15, West Conshohocken, PA: ASTM International.

Atlee, J. and T. Roberts, 2007. Cradle to cradle certification: A peek inside MBDC's black box. Environmental Building News, February 1, 2007.

Bernier, P., R.A. Fenner and C. Ainger, 2010. Assessing the sustainability merits of retrofitting existing houses. *Proceedings ICE Engineering Sustainability*, 63(4): 187–201.

Brandi, H.S., R.J. Daroda and T.L. Souza, 2011. Standardization: An important tool in transforming biofuels into a commodity. *Clean Technology and Environmental Policy*, 13: 647–649.

Canadian Green Building Council, 2018. LEED in Canada Education Promotes Green Building, leedcanada.net. Accessed June 29, 2018.

CEEQUAL, 2016. Section 2: Project Management. www.ceequal.com/version-5/section-2-project-management/. Accessed July 8, 2018.

Chin, K., D. Schuster, D. Tanzil, B. Beloff and C. Cobb, 2015. Sustainability trends in the chemical industry, *CEP*, 111(1): 36–40.

Chisholm, D., K. Reddy and M.R.O. Beiler, 2017. Sustainable project rating systems, including envision. In: *Engineering for Sustainable Communities, Principles and Practices*, W.E. Kelly, B. Luke and R.N. Wright (eds), Reston, VA: ASCE Press, pp. 307–326.

Ding, D.K.C., 2008. Sustainable construction-the role of environmental assessment tools. *Journal of Environmental Management*, 86: 451–464.

Elefsiniotis, P. and D.G.J. Wareham, 2005. ISO 14000 environmental management standards: Their relation to sustainability. *Journal of Professional Issues in Engineering Education and Practice*, 131: 208–212.

Elkington, J., 1998. *Cannibals with Forks: The Triple Bottom Line of 21st Century Business*, Gabriola Island, BC: New Society Publishers.

Feldman, I.R., 2012. ISO standards, environmental management systems, and ecosystem services. *Environmental Quality Management*, 21: 69–78.

Ghumra, S., M. Watkins, P. Philips, J. Glass, M.W. Frost and J. Anderson, 2009. Developing an LCA-based tool for infrastructure projects. In: *Proceedings of the 35th Annual Conference of Association of Researchers in Construction Management*, A.R.J. Dainty (ed), Nottingham: ARCOM, pp. 1003–1010.

GPM (Green Project Manufacturing), 2012. GPM P5 Standard for Sustainability in Project Management) Management 1.5. https://greenprojectmanagement.org/the-p5-standard. Accessed July 8, 2018.

GRI (Global Reporting Initiative), 2016. About Sustainability Reporting. www.globalreporting.org/information/sustainability-reporting/Pages/default.aspx. Accessed November 13, 2018.

ISCA (Infrastructure Sustainability Council of Australia, 2018. Tools and Resources. www.isca.org.au/tools_and_resources. Accessed June 30, 2018.

ISI (Institute for Sustainable Infrastructure), 2015. Envision Rating System for Sustainable Infrastructure, Washington, DC. https://sustainableinfrastructure.org/envision/. Accessed July 8, 2018.

ISI, 2018a. Envision: How it Works…. http://sustainableinfrastructure.org/envision/how-it-works/. Accessed June 25, 2018.

ISI, 2018b. Envision Sustainability Professional (ENVSP). http://sustainableinfrastructure.org/env-sp-directory/envision-sustainability-professionals/. Accessed June 25, 2018.

ISO, 2006. Three Pillars of Sustainable Development-Metrology, Standardization and Conformity Assessment. www.iso.org/files/live/sites/isoorg/files/archive/pdf/en/devt_3pillars_2006.pdf. Accessed July 8, 2018.

Manitoba Hydro, 2000. Sustainable Development Report, Winnipeg, MB.

MOEJ (Ministry of the Environment, Japan), 2009. Annual Report on the Environment, the Sound Material-Cycle Society and the Biodiversity in Japan. www.env.go.jp/en/wpaper/2013/pdf/00_cover.pdf.

Sikdar, S.K., D. Sengupta and R. Mukherjee, 2017. *Measuring Progress towards Sustainability: A Treatise for Engineers*, Cham, Switzerland: Springer International Publishing.

Takakuwa, S., 2013. A perspective on manufacturing and environmental management, Chapter 9. In: *DAAAM International Scientific Book*, B. Katalinic and Z. Tekic (eds), Vienna: DAAAM International, pp. 213–234. doi:10.2507/daaam/scibook.2013.09. Accessed July 8, 2018.

UNGC (United Nations Global Compact), 2015a. The Ten Principles of the UN Global Compact. www.unglobalcompact.org/what-is-gc/mission/principles. Accessed July 8, 2018.

UNGC, 2015b. United Nations Global Compact. www.unglobalcompact.org/. Accessed July 8, 2018.

USACE (United States Army Corp of Engineers), 2013. Biddability, Constructability, Operability, Environmental and Sustainability (BCOES) Review. ER 415-1-11. Washington, DC: USACE. www.publications.usace.army.mil/Portals/76/Publications/EngineerRegulations/ER_415-1-11.pdf?ver=2014-07-23-093636-257. Accessed July 8, 2018.

USGBC (U.S. Green Building Council), 2018. LEED is Green Building. https://new.usgbc.org/leed. Accessed July 8, 2018.

5

Sustainable Material, Energy, and Water Use

5.1 Introduction

Engineers are employed in many different industries and each of these has their own activities, such as mining, food processing, forestry, manufacturing, energy, and service industries. These industries are essential to the economy and support of society. However, their activities can conflict with the goals of a sustainable environment. Some of these aspects in the production processes include (a) the use of renewable and nonrenewable materials, (b) use of nonrenewable natural energy resources, and (c) the emissions of atmospheric, liquid, and solid wastes from these industries that can impact human health and the environment (Figure 5.1).

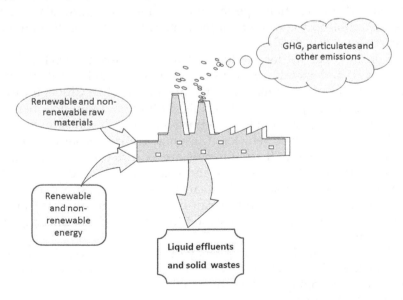

FIGURE 5.1
Interactions between manufacturing industries and the environment.

In summary, sustainable engineering practices should aim to (a) use more efficiently renewable and nonrenewable natural resources to allow future generations to continue to benefit from these resources and (b) reduce the discharge of waste materials into the environment to minimize the impact on humans and the environment. The discussion in this chapter concerns the role of engineering in providing more sustainable processes.

5.2 Material Selection and Use

Engineers are involved in the processes for the extraction of natural resources. Better processes can be designed for more efficient extraction of minerals and petroleum that have a lesser impact on the environment. Both minerals and petroleum are deemed nonrenewable as once they are extracted, the process is irreversible. Metals and mineral resources can be recycled, enhancing their sustainability. Social acceptance of mining has been low due to numerous instances of environmental impacts, profit-sharing problems, inequity, and land use issues. More standards are needed for addressing local concerns.

Some of the environmental impacts are related to contamination due to arsenic and other metals; acid mine drainage and groundwater contamination are shown in Figure 5.2. Storage of waste rocks and tailings is a major problem due to dam failures and release of metals from the stored tailings via oxidation and precipitation. Speciation of the metals is particularly important as this controls the mobility of the metals. For example, chromium(VI) is much more toxic than Cr(III).

Most of the materials for products from the lithosphere, such as stones, sand, gravel, and metals, are considered nonrenewable as they are not replenished within the human lifetime. In countries like China, the extraction rate is increasing. Rare-earth metals, gallium and indium, are needed for solar panels, neodymium for wind turbines, and dysprosium for batteries in hybrid and electric vehicles. These metals are becoming exhausted (Heinberg 2011). Some of these minerals are extracted using child labor in some countries. In addition, energy and water are required for the extraction processes. Various approaches can be used to reduce material use. Material thicknesses in a product can be reduced to minimize material requirements. Products can be designed to last longer that are upgradable or easy to repair.

The Environmental Protection Agency (EPA) in the United States and the World Business Council for Sustainable Development have developed various tools for developing products or functions, or producing them in alternative ways. Design for environment (DFE) or ecodesign implementation can assist this process as mentioned previously in Chapter 2. Metrics are

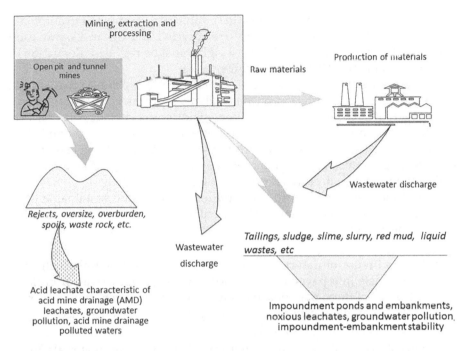

Mining, extraction and processing

Open pit and tunnel mines

Raw materials

Production of materials

Wastewater discharge

Rejects, oversize, overburden, spoils, waste rock, etc.

Wastewater discharge

Tailings, sludge, slime, slurry, red mud, liquid wastes, etc

Acid leachate characteristic of acid mine drainage (AMD) leachates, groundwater pollution, acid mine drainage polluted waters

Impoundment ponds and embankments, noxious leachates, groundwater pollution, impoundment-embankment stability

FIGURE 5.2
Aspects associated with mining and natural resource extraction.

chosen for evaluation, comparisons are made of alternatives, and implementation and further measurements are done to continue improvement. This approach needs to be done over the life cycle not just the production phase so that the most appropriate raw materials can be chosen and used through the production phase to disposal. Criteria can be based on their ability to be reused and use of less material.

Choosing low toxicity materials is another important aspect over the life cycle but this can be complicated. The product itself must not contain toxins such as lead or pesticides. But also, the process for isolating or producing the material must not use extensive amounts of pesticides or other toxic materials such as cyanide. Materials can also be chosen based on lower carbon dioxide emissions during production. The 12 principles of green chemistry that were previously discussed should be followed for selection of materials. For example, renewable raw materials should be used as much as possible. Materials should be chosen to avoid hazard and risk in the environment and should be as biodegradable as possible so they do not remain in the environment.

Products and packaging designed for reuse or disassembly can reduce overall material use. Products that are repairable will also have a longer life. Modularity also can enable easy upgrade in the products. Some companies take back materials so that they can be reused in another or similar product.

For example, in the European Union (EU), vehicles are returned to automakers, disassembled, and the materials are recycled (European Commission 2018). A total of 85% of the vehicle by weight must be recycled. Reduction of the number of materials facilitated this process.

Transportation is another part of the life cycle of a product that should be considered. Transportation of materials and products produces greenhouse gases (GHGs) and other emissions. Shortening distances is one way to reduce the emissions and fuel requirements. One way to accomplish is through the use of local materials. Products can be designed that take up less space through reduced packaging or packaging more tightly. Another way is to design for more concentrated products with lesser water.

Geotechnical engineering, for example, can incorporate sustainable practices into materials management. Basu et al. (2015) and Puppala et al. (2018) listed some of these:

- Use and reuse of natural materials, and/or waste materials such as C&D waste into pavement subgrade, earthwork materials, or backfill. Some of these materials include fly ash, bottom ash, rubber mix, asphalt shingles, concrete, asphalt pavement blast furnace slag, shredded tires, and recycled glass.

- Biostabilization processes or natural fibers or waste materials for soil reinforcement. Biocementing is a new process (Fukue et al. 2011, 2013). Vegetation growth has been promoted. Geosynthetics such as geogrids and geotextiles have been found to have less embodied energy (Jones and Dixon 2011). Recently, geogrids can be made from recycled plastic bottles (Puppala et al. 2018).

- Reuse of foundations and other structures; Butcher et al. (2006) found that embodied energy was half that of a new foundation.

- Beneficial use of underground space for facilities such as water systems, pedestrians, storage of waste, or carbon dioxide to conserve land use.

- Recovery of minerals and exploitation of renewable resources such as geothermal energy. Some researchers from the University of Texas at Arlington (UTA) are examining geothermal energy for melting snow and ice from roads (Puppala et al. 2018).

- Redevelopment of brownfields (to be discussed in the next chapter).

- Sustainable land use.

- Remote sensing (Light detection and ranging, satellites, and unmanned vehicles) to monitor and manage infrastructure to ensure resilience.

Biocementation is the microbially induced production of carbonates similar to that of natural diagenesis. It is useful for strengthening soil (ground improvement) (Fukue et al. 2011). One of the most efficient techniques is to utilize ureolytic microbes that produce the enzyme urease. Ureolytic bacteria are grown in an appropriate medium and mixed with the reactive solution consisting of $CaCl_2$, $MgCl_2$, NaCl, buffer materials, etc., before injecting into the soil.

Fukue et al. (2011) isolated a strong ureolytic microbe (NO-A10) from a Japanese soil and used the microbes to improve the strength of sandy soils through microbially produced calcite ($CaCO_3$). In the reaction, the urease enzyme produced by ureolytic bacteria can be used to induce hydrolysis of urea ($(NH_2)_2CO$), as follows:

$$(NH_2)_2 CO + 2H_2O \rightarrow CO_3^{2-} + 2NH_4^+ \tag{5.1}$$

The reaction is followed by the production of carbonate.

$$\frac{1}{2}Ca^{2+} + \frac{1}{2}Mg^{2+} + CO_3^{2-} \rightarrow \frac{1}{2}CaMg(CO_3)_2 \downarrow \tag{5.2}$$

Soft soils are subject to geo-disasters due to high compressibility, low strength, low bearing capacity, and high liquefaction potential. The biocementation process is initiated by injecting the solution of microbes and reactive agents into fine soils. Studies on sandy soils cemented with carbonates induced by the microbes have shown significant increases in unconfined compressive strengths attributable to carbonate content (Fukue et al. 2013). The gain in strength can be governed by the numbers of injections, and the carbonate content obtained can be predicted from the concentrations of agents and the number of injection times. The microbes must be well distributed into the target area.

Land resources have been used for many applications. Some of these are foundations for infrastructure, agricultural and forestry purposes, resource extraction, and recreation. Natural resource assessment is performed to evaluate and improve the condition. The assessment can be for water, air, ecology, and soil. Wright (2017) has indicated that:

- Sustainable land use involves viewing the ecological system as valuable and a finite resource.
- Natural and ecological systems have economic values.
- Climate change exists as shown by rising sea levels, melting ice, and rising temperatures.
- Physical human resources are linked to the health of natural resources.
- Design must be forward thinking instead of backward due to changing conditions.
- A more holistic approach is necessary for sustainability in projects.

Processing of the mineral resources can be improved by more efficient extraction of the metal and minerals by hydrometallurgy and wet processing (Calas 2017). Recycling and reuse are needed to increase the use of the materials and reduce wasted resources. This must also include recycling of cell phones, computers, etc. that may contain small quantities of the metals. In the EU, glass is recycled at a rate of 70%–90% in some countries. However, quality decreases and thus reuse, as building aggregates is common. Aluminum recycling rates have increased globally from 19% in the 1950s to 36% in 2014 and will increase above 40%. Some metals (such as To, Cr. Mn, Fe, Cu, Ni, Zn) are recycled at rates greater than 50% while other rates are at 1%–10% (Sb and Hg) or less than 1% (vanadium, zirconium) (UNEP 2011).

Research by academia and industry is needed to better use minerals and reduce environmental and societal impacts. In particular, low-grade ores and wastes should be reprocessed to extract minerals and reduce acid mine drainage and other pollution. De Villiers (2017) has proposed methods to enhance extraction of minerals that will increase mine life and profitability. As the flotation/milling step is 80%–90% efficient, there is room for improvement. Improved by-product extraction such as copper and nickel, can improve process economics. Two-stage hydrometallurgical processes and solvent extraction/electrowinning have been very beneficial for copper mining of ores of low grade (Bartos 2002). Bio-hydrometallurgical processes have enhanced copper and gold recovery. Iron-ore sintering is also increasing in importance and increasing the use of low-grade ores. Quantification of energy and materials requirements, process efficiency, and emission and wastes produced can be performed.

5.3 Energy Selection and Use

Energy is used for electricity, transportation, heat, and many other uses. The sources of energy are usually classified into (a) nonrenewable and (b) renewable. The main nonrenewable source of energy is fossil fuels (hydrocarbons and coal) while another source of nonrenewable energy is uranium for the production of nuclear energy. Renewable energy sources refer to sustainable sources such as solar, wind, tidal, wave, geothermal, and biomass.

5.3.1 Fossil Fuels

Due to the implementation of hydraulic fracturing, the United States has shifted to be an oil and natural gas producer. Natural gas plants emit half the carbon dioxide of coal based for the same power (Sikdar et al. 2017). This is due to the lower carbon intensity of natural gas (13.7 g C/MJ) compared to coal (24.4 g C/MJ) (Rubin 2001). Therefore, using natural gas is beneficial. However, there are potential environmental issues related to hydraulic fracturing.

Hydraulic fracturing is a process designed to remove tight oil and gas from petroleum-rich geologic formations by injecting proppants (sand) and a mixture of chemicals and other substances at high pressure. The chemicals are used to maintain the openings of the fractures, inhibit bacterial growth, and efficiently deliver the fluids. Investigations have reported the contamination of water supply wells due to chemicals that are injected to create hydraulic fractures in oil and gas reservoirs and uncontrolled oil and gas migration resulting in water contamination with salts, organic compounds, metals, metalloids, other inorganics, aromatic carcinogens, toxins, biocides, and surfactants.

5.3.2 Nuclear Energy

While nuclear energy is a clean energy and does not produce GHG emissions, nuclear energy production is particularly related to the safety issues during production and the problem of disposal of high-level radioactive waste. Yong et al. (2014) indicated that the following aspects are of particular importance: (a) the level of radioactivity of the spent fuel (whether it is high-, medium-, or low-level radioactive waste), (b) the time necessary for the heat to dissipate by the spent fuel, (c) the time necessary to reach the required level of radioactivity of the spent fuel to reach acceptable limits. Yong et al. (2010) discussed many of the aspects related to waste disposal. Many countries use nuclear power. However, in the past there have been some significant accidents (Fukushima, Chernobyl, and Three Mile Island). Radioactive waste generation is also a major problem as adequate facilities are difficult to obtain. More recent developments in breeder reactors enable more efficient energy production and thus could be a sustainable energy source if safety is ensured.

Proper emergency planning and remediation schemes for an unexpected disaster need to be developed for events such as the fall-out of radioactive cesium from the destruction of the nuclear power plants at the Fukushima Daiichi Nuclear Power Plant in Ōkuma, Fukushima Prefecture. A tsunami occurred following the Tōhoku earthquake on March 11, 2011. Although the active reactors were automatically shut down after the earthquake, emergency generators were damaged by the tsunami. These were needed to operate the pumps to cool the reactors. The problems with the cooling led to the meltdown of three reactors releasing radioactive material. Later, the Fukushima Nuclear Accident Independent Investigation Commission (NAIIC) concluded that Tokyo Electric Power Company (TEPCO), the plant operator, did not prepare proper risk assessments and develop evacuation plans and thus many of the impacts could have been avoided.

5.3.3 Renewable Energy

Alternatives to traditional nonrenewable fossil fuel energy sources such as coal and hydrocarbon resources are increasingly desired due to (a) generation of GHGs that can directly contribute to global warming and (b) the need to replace nonrenewable resources with renewable sources. Global warming concerns are tightly connected with fuel energy due to GHG production (carbon dioxide, nitrous oxide, and others). Carbon dioxide is produced during transportation, generation of electricity, industrial and building operations, and changes in land use. Some of these renewable sources include hydroelectric, solar, geothermal, biomass, wind, and tidal and wave. In reality, solar, geothermal, tidal and current and wind should be considered as non-depleting energy resources over any given time period.

Solar and wind energy are used for electrical power. Solar photovoltaic panels contain solar cells that produce about 0.5 V each that are connected in modules. DC power is produced. An inverter is needed to convert to AC power. Costs of the panels have decreased substantially recently but they are still expensive. No emissions are produced but chemicals, energy, and materials are required for production of the panels. The panels are particularly useful for off-grid applications. Photovoltaic capture and conversion rates will need to be improved. Solar energy can also be in the form of thermal energy. Water can be heated for steam boilers.

Wind turbines consist mainly of three blades on a horizontal axis on a tower about 60–90 m high. Strong and reliable winds are needed. They can be in a wind farm or individually. Wildlife such as birds and bats needs to be protected from the blades, so appropriate siting is essential. Aesthetics are also a concern from many residents so offshore is an option used but this can be expensive for the transmission lines. Noise is another aspect that has generated complaints. Improved design has reduced this problem but low frequency sounds may be an issue.

One of the major impacts of alternative fuels is land use. Solar and wind farms require substantial amounts of land. Biofuels require land to be tilled. Figure 5.3 shows the comparisons (Cheng and Hammond 2017). Nuclear is clearly the least land intensive whereas biomass is the most. This is due to the cultivation, harvesting, and transportation requirements. Renewable energy is clearly a dilute form of energy. However, there are many other considerations. Geothermal energy extracts heat from volcanic activity. Wells are drilled to capture the steam, which then can be used to turn steam turbines and generate electricity. In other regions, geothermal energy can also be extracted but much deeper wells are needed.

Hydroelectricity is used in many countries. Large lakes are produced by damming a river. Potential energy from falling water is then captured to turn turbines. However, there are some aspects of hydroelectricity that cause impacts. Often people need to be moved due to flooding such as for the Three Gorges Dam in China and the Aswan Dam on the Nile River.

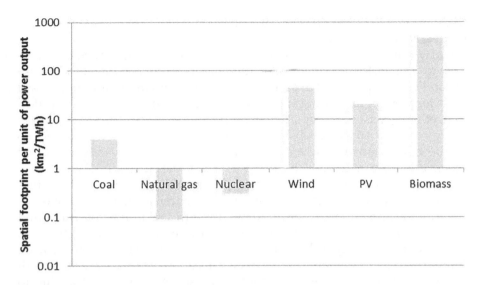

FIGURE 5.3
Spatial footprint per unit of energy output for various energy sources. (Data from Cheng and Hammond [2017]; Gagnon et al. [2012].)

In addition, the ecology of the area is changed and fish migration is hindered. It is carbon-neutral and tends to be less expensive than fossil-based power. Silt deposition reduces the lake volume and the power generated over time. Organic matter degradation produces methane, a GHG. It is, however, a constant and reliable source of energy. Micro-hydropower could be useful for small businesses, farms and houses as up to 100 kW can be generated without a reservoir.

Biodiesel is being used more frequently for transportation via transesterification of triglycerides of fatty acids to produce alkylated fatty acid. It can be produced from various oils including mustard, palm, rapeseed (canola), soybeans, and sunflower (Weeks 2005). The by-product glycerol needs a high-value use. Algae biodiesel is being developed. However, there are challenges as enclosed bioreactors are needed to avoid microbial contamination and substantial amounts of water are needed, which will be expensive to remove.

Soybean and rapeseed are the most common sources of biodiesel in the United States and Europe, respectively. Water frying oil has also been used. Compared to diesel, biodiesel fuel use produces less wastewater and emissions. It is believed to be more biodegradable. However, this may not always be the case. Tests performed by Saborimanesh and Mulligan (2015) indicated that in an oil spill scenario, the highest level of biodegradation of diesel (27%), biodiesel (28%), and light crude oil (30%) during the treatment without any additives, and 28% (diesel), 28% (biodiesel), and 30%

(light crude oil) in the presence of a biosurfactant in the first 7 days of the biodegradation period.

Biofuels in Brazil, in particular, are produced from sugar such as sugarcane to ethanol. In the United States, there has been opposition due to the use of food corn to produce ethanol. Large amounts of land that require fossil fuels for plowing and tilling and fertilizer addition are problematic. In addition, ethanol must be transported by petroleum. Ethanol is volatile and thus energy can be lost. Often subsidies are used, such as in the United States, so the benefits of ethanol use are questionable. Biofuels can also be produced from various wastes.

Traditional biomass has been used for heat and power but recently bioethanol and biodiesel are being produced for transportation. Biomass in the form of agricultural waste, municipal solids, saw mill wood waste and forest wastes can be anaerobically digested to produce methane that can be used for electricity or heat. The systems are modular and can be used industrially and for communities. Biomass can also be burned. Although it releases carbon dioxide, it is considered to be carbon neutral due to removal of carbon by plants (Randolph 2008). Burning releases particulate matter that leads to smog or release of carcinogenic matter. Methane emissions can also be captured from landfills to produce electricity or a fuel for equipment.

A newer renewable source of energy that is being developed is called osmotic power or salinity gradient energy and involves fresh and salt water (Achilli and Childress 2010). Although osmotic power and hydropower are similar as both use hydroturbines, they differ as a hydropower plant exploits the energy generated by river water with a dam and pressure-retarded osmosis (PRO) generates osmotic power based on the salinity gradients via a membrane-based process. Water from a low osmotic pressure feed solution goes toward a high osmotic pressure draw solution due to hydraulic pressure (She et al. 2013). A back pressure is applied on the high salinity draw solution side to retard the permeate water flow in the PRO process. Power can then be produced by releasing the pressure from the solution through a turbine (Kim and Elimelech 2013).

Achilli and Childress (2010) reviewed PRO technology that was invented by Prof. Sidney Leob and has attracted increasing attention. The first PRO osmotic power plant was built in 2009 by Statkraft Company in Norway. Others working on the osmotic energy process include WETSUS in the Netherlands and in Japan, Kyowakiden Industry Co. in collaboration with Kyushu University, Nagasaki University, Tokyo Institute of Technology, and Mega-ton Water System. The latter produces with heir PRO prototype plant $9\,W/m^2$. Projects have been started/are being developed in different countries around the world: Italy (UNIPA), Israel (IDE), Korea (GMPV), Colombia, and the River Thames in UK. In Canada for northern communities there are particular advantages for off-grid communities where polluting, diesel generators are mainly employed. In addition to scale-up, other research is taking place with regard to reducing the potential for membrane fouling (Abbasi-Garravand et al. 2016, 2017).

5.3.4 Carbon Capture and Storage

Carbon capture and storage (CCS) is being used to store carbon dioxide in geological setting for coal-based power plants to reduce emissions (Saskpower 2018). In 2014, the Boundary Dam Power Station near Estevan, Saskatchewan, became the first to successfully use CCS technology. It produces 115 megawatts (MW) of power, a sufficient amount for about 100,000 Saskatchewan homes. It is also able to reduce SO_2 and CO_2 emissions by up to 100% and 90%, respectively. There are disadvantages to the technology. CCS is expensive and requires substantial energy. There could be accidental releases of carbon dioxide. The geology may not be appropriate in all areas and there could be chemical reactions that are not expected due to a lack of understanding.

5.3.5 Energy Efficiency

Since increased energy efficiency will lead to reduced carbon dioxide emissions, this will reduce global warming. According to Sikdar et al. (2017), some methods to achieve this include:

- Wasting of energy through flaring at energy production facilities should be reduced. Usually this is done, as it is the most economical option.
- Increasing energy efficiency and generation. Natural gas plants (45%) are more efficient than coal ones (33%). Combined cycle power plants are more efficient as they can recover heat in exchange for power. Cogeneration or combined heat and power is another approach to produce power and use the exhaust gases for heating. Integrated gasified combined cycle produced steam and syngas from fossil fuel or biomass, which is then used to produce electricity. The steam from these stages is used for power generation and achieving 60% energy efficiency. Thus, innovations can lead to improvements.
- As coal is still the most widely used power source worldwide (approximately 40% of the world's electricity), carbon capture and storage (CCS) has been developed to capture carbon dioxide from coal plants and storing it in a geological formation. The carbon dioxide is purified and compressed.

Approximately 61% of the energy produced is lost due to poor insulation, car gasoline efficiency, and power plant operation (Sikdar et al. 2017). Enhanced energy sustainability can be obtained by improvement in energy efficiency at all levels, reduction in energy use, and replacement of fossil fuels by renewables. Policy development to promote renewable energy use, reduce overall energy use, and increase energy efficiency is also required.

Energy efficiency in buildings is important as it makes up a large portion of energy consumption. Green building design concentrates on new buildings only but older buildings are more common (50:1) (Landsberg and Lord 2009). Energy audits can identify energy uses and thus potential areas for reduction. Lighting systems can also be more efficient. Combining heat and power is an option also for industrial processes. Changing coal or oil to natural gas can reduce GHG emissions by half. Methane releases must be minimized, however, since it is a higher global warming potential (GWP) than carbon dioxide.

Manufacturing processes can be designed for energy efficiency. A life cycle assessment (LCA) can be performed to analyze the energy and embodied energy at all stages of the life cycle. Recycling and reuse of materials can reduce energy requirements by 50% compared to virgin materials. Use of ambient temperatures and pressures for processing can reduce energy requirements. Design for environment and industrial ecology approaches can also reduce wastes.

5.3.6 Energy Source Selection

When selecting a renewable energy source, the location and environmental conditions must be considered. Subsidies from the government will be required to enable economic competitiveness. Higher energy efficiency, combining heat and power and reducing energy use are key strategies. Building design, material selection, process design, recycling and reducing water use all can reduce energy requirements.

Annual death and man-days lost can be used as indicators for safety. Coal and oil production have led to lung diseases, explosions, and ecological damage. Alternative fuels are more safe, but other issues related to continuous energy production lower energy density. Therefore, indicators for consideration of energy selection can include:

- Accidents
- Economic viability
- Energy density
- GWP
- Health impacts
- Land use changes
- Supply chain issues
- Transmission or transportation costs
- Water used

For evaluating the sustainability of a technology, different indicators are used. Mata et al. (2011) evaluated diesel versus biodiesels. The indicators include life cycle energy efficiency (total energy content to total energy used), fuel efficiency ratio (energy to fossil energy used), land use intensity, carbon footprint, and embodied energy. The results of the analysis are shown in Table 5.1. Some biodiesels are more favorable in fuel efficiency than fossil diesel. However, carbon footprint can be higher as in the case of microalgae biodiesel. It is, however, much less land intensive ($0.9\,m^2$ year/kg biodiesel) than soybean ($20.3\,m^2$ year/kg biodiesel) and sunflower biodiesels ($12.0\,m^2$ year/kg biodiesel).

American Society for Testing and Materials (ASTM 2010) has developed a guide for reduction of GHG for businesses and industry (Figure 5.4). It is a three-tier approach. The first tier uses measures that can be implemented quickly and includes conservation and efficiency of energy. Direct, and indirect measures such as reduction in transportation or water use can be implemented. Various tools such as GHG calculators can be used to track progress. The second tier can be undertaken as the next step and includes implementation of new technologies for GHG reduction. These can include use of solar panels, wind turbines, geothermal, green roofs, restoration of vegetation, or purchase of energy from an electrical power plant. Tier III involves a long-term approach with developing technologies such as carbon sequestration or hydrogen fuel. The purpose of the tiered approach is iterative for long-term planning.

In summary, energy systems should be sustainable economically, environmentally, and socially. A life cycle approach should be employed to evaluate the most appropriate choice. They need to be reliable, affordable, and safe. Smart grids will assist in their optimization. Efficiency should be maximized to conserve resources and environmental degradation from fossil fuel emissions and spills need to be minimized. Importing fuels should be minimized and the life of energy systems should be maximized. Some indicators (Wright 2017) for energy system sustainability include the ratio of end use of energy to energy generated, GHG generation, percentage of energy from fossil fuels, energy lost or wasted, and energy usage.

TABLE 5.1

Comparison of Fossil Diesel with Selected Biodiesels (Mata et al. 2011)

Indicator	Fossil Biodiesel	Soybean Biodiesel	Microalgae Biodiesel	Sunflower Biodiesel
Life cycle energy efficiency[a]	6.3	0.4	1.8	1.0
Fuel efficiency ratio[b]	6.3	0.4	0.6	1.0
Carbon footprint ($kgCO_2$ eq./kg fuel)	4.3	2.9	5.4	1.9

[a] Total energy content/total energy used.
[b] Ratio energy obtained/fossil energy used.

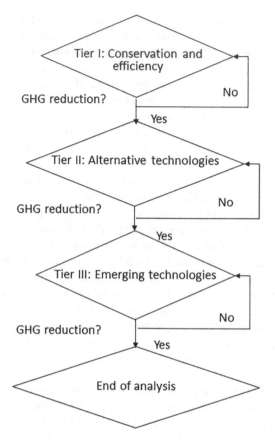

FIGURE 5.4
Three-tier approach for GHG reduction. (Adapted from ASTM [2010].)

5.4 Water Use and Reuse

Water is a resource required for many purposes such as drinking water, agriculture, domestic and industrial uses, transportation, recreation, power production, and support for biota. Increasing water shortages are due to (a) increased demand in excess of supply, (b) aquifer depletion, (c) lack of rain and other forms of precipitation, (d) watershed and water resource mismanagement, and (e) other aspects related to climate change.

The importance of water has been highlighted in the Johannesburg World Summit on Sustainable Development (WSSD) in 2002. "Water and Sanitation" was identified as one of their five thematic areas at WSSD and more recently in Rio Summit +20 in 2012, as discussed in Chapter 1. This section will examine elements required toward more sustainable water management. According to Yong et al. (2014), *sustainable water management* is defined as "all the activities associated with usage of water in support of human needs and aspirations, must not compromise or reduce the chances of future generations to exploit the same resource base to obtain similar or greater levels of yield."

Sufficient water quality and quantity are essential. The first Dublin–Rio principle has emphasized the need for sustainable water resource practices. It states that: "Fresh water is a finite and vulnerable resource, essential to sustain life, development and the environment."

Depletion of aquifers occurs when the rainfall is insufficient, excessive groundwater extraction, and/or aquifer recharge is exceedingly slow. Continued water use from that aquifer is thus not sustainable. In regions of excessive groundwater extraction from interlayered aquifers with soft aquitards and aquifers, subsidence of the ground surface can occur, resulting in flooding or coastal land surfaces that can reach levels below sea level. In the latter case, containment dikes must be built to avoid seawater intrusion, local flooding, and contamination of the aquifers. Foundation damage can occur in buildings and other structures.

To obtain sustainable water usage, water use will need to be employed more efficiently, and reuse will need to be increased. In developing countries, water supply can be scarce and of low quality, and thus water management practices must be improved. One method is to increase the permeability of paved surfaces. Runoff water can pick up contaminants on roads and parking lots. This runoff reaches surface waters, providing a source of contamination. Kabbes et al. (2017) have indicated that the use of plants in green infrastructure or low impact development (LID) can reduce runoff and impact water quality and allow groundwater recharge.

With regard to climate change, increases in rainfall intensity, storms, and droughts must be considered in design of infrastructure to mitigate damage. Changes in 25- to 50-year design lives must be considered by engineers. Some guidance is available from the U.S. Bureau of Reclamation for climate change scenarios for reservoir design (USBR 2016). The National Oceanic and Atmospheric Administration (NOAA 2012) provides guidance for sea level rise in coastal regions.

In developed countries, a combination of regulations and initiatives such as water metering is needed to reduce water use. Infrastructure must also be maintained on a regular basis to reduce leakage from water supply systems. Flooding is experienced by many cities worldwide. Upstream dams instead of proactive policies for water sustainability can lead to water supply problems for downstream cities. Reuse of wastewater can reduce use of fresh water. Policies such as introduction of taxes or tax credits may also be beneficial for water use reduction. At the building level, other approaches include the use of green roofs to adsorb roof runoff, and allowing rain water infiltration through permeable pavements to return to the groundwater (König 1999).

Drinking water quality is not required for all purposes such as cooling water, washing, toilets, and many industrial uses. Different treatment levels can be used for various purposes. Presently, wastewater management systems act as a foundation for modern public health and environment protection. The idea of more suitable wastewater management systems is to use less energy, allow for elimination, or beneficial reuse of biosolids, restore natural nutrient cycles, have much smaller footprints, be more energy efficient, and design to eliminate exposed wastewater surfaces, odors and hazardous by-products (Daigger and Crawford 2005). In addition to the technical aspects of a wastewater treatment technology, selection of a particular technology should be based on all aspects that determine its sustainability. More information on this is provided in Chapter 6.

Water footprint is an indicator that is used for direct and indirect water use of blue, green, and gray water (Hoekstra and Chapagain 2007). Blue water includes surface and groundwater. Green water is that which is stored in the soil for use by plants. Gray water is polluted and is measured as the volume of water needed to reach water quality standards. Water is not destroyed and is generally renewable through the hydrologic cycle but it can become polluted and extracted faster than it can be replenished in certain areas. It may also be not available where it is needed. California, for example, has been restricting water use due to shortages. A total of 80% of the demand is where only a few centimeters of rain are received per year.

Large amounts of water are needed in industries, such as for car manufacturing (148,000 L/car) (Robertson 2017), paper (2 L/g), and microchips (16 L/g) (Allenby 2012). The water footprint is expressed typically in water volume per unit of product. A water sustainability tool has been developed by the Global Environmental Management Initiative (GEMI 2015) for businesses. The tool includes five steps with key questions and outputs. The steps are:

- Water use, impact, and source assessment
- Business risk assessment
- Business opportunity assessment
- Strategic direction and goal setting
- Strategic development and implementation

Practical approaches for water management include:

- Consideration and engagement of other stakeholders for management of water resources
- Reduction of water use (reduce leakages, recycle water for cooling/heating, and other water uses)
- Selecting appropriate water quality for its use
- Minimizing impact on water quality particularly by reducing pollutant introduction into the water instead of water treatment after being contaminated
- Promoting water sustainability and stewardship

Cooling systems are often single pass for water use. Water is often discharged after absorbing the heat. Corporate culture may make change difficult. Water infrastructure systems must also be improved to enhance water reuse, recovery of resources from biosolids and nutrients, and reduced energy use. Climate change is contributing to future water uncertainty. Fresh water is becoming more polluted due to urban growth.

Rainwater harvesting involves collecting and storing rain (stormwater) for later use. Rain gardens, porous pavement, bioswales, and green roofs are used to clean the stormwater. The system usually includes the catchment area, pipes or gutters for transporting the water, filters for removal of debris, and a storage tank (cistern). If the water is to be used for drinking then further treatment will be necessary. Otherwise, the water can be used for cooling, toilets, irrigation, or fire sprinkler systems. Design considerations will include rainfall estimates and water uses. Storage will be based on the time when demand exceeds supply.

Gray water, water collected from bathroom sinks, air conditioning condensates, drain water, cooling tower makeup, showers, tubs, and washers can reduce water treatment requirements. A system, however, will be needed to keep potable and nonpotable water separate. Black water from toilets, kitchen sinks and dishwater cannot be mixed with gray water and has to be removed from the building but could be treated onsite.

Stormwater management in a more sustainable way involves maximization of infiltration. This allows recharging of groundwater and removal of pollutants by filtration. Three strategies can be used, focus on small storms, retain the water at the source, and use permeable surfaces as much as possible. These systems should be natural and of low impact.

The U.S. Clean Water Act indicates that best management practices for reducing pollutant discharges into lakes and rivers should be used. These include low-impact development (LID) such as constructed wetlands, rain gardens, vegetated swales, porous pavements, and green roofs.

Rain gardens are basins used for stormwater runoff collection for infiltration into the soil below, usually within 72 hours. Swales are vegetated drainage channels for capturing and transporting runoff. They reduce the velocity and pollutants from the stormwater and allow infiltration and groundwater recharge. Native species vegetated swales are called bioswales. The residence time in the bioswales is typically 1–4 days. This allows the sediments to settle but does not inhibit mosquito growth. Vegetation can be a component of an impervious parking area. Porous pavement allows water to infiltrate but still provide the ability to walk or drive on it. Green roofs are composed of a waterproof layer, a medium for plant growth and the plants. They can absorb roof runoff, store carbon, reduce heat buildup, and lower energy requirements. Minimization of water use, reduction of water leakage, and use of recycled water and rainwater (APA 2016) are sustainable practices.

5.4.1 Water–Energy Nexus

Water is linked to energy production. Water can be used for thermoelectric power and geothermal energy. About 45% of water withdrawal is for thermoelectric power in the United States (Maupin et al. 2014). Recirculation of cooled water in the western United States decreases water consumption. The geothermal sites are mainly in volcanic and seismic regions in countries such as Iceland, Spain, France, Hungary, Japan, Mexico, Russia, and the United States (in the states of California and Hawaii) (Chamley 2003). Either the steam of geysers or very hot liquids of 100°C or more, or groundwater at a depth of 100 m that is about 15°C can be used for energy. Some challenges regarding geothermal energy include (a) loss of heat from transportation of water and (b) potential inducement of earthquakes. The potential benefits are that negligible amounts of contamination are produced, and there is little or no production of GHGs, or waste production. Water is used for cooling of most energy production forms with the exception of wind, ocean, hydro, or osmotic power. For enhanced oil recovery, three barrels of water are needed per barrel of oil. Fracking also requires substantial water use. Nuclear energy and biofuels also require substantial amounts of water. Water recycling and reuse needs to be practiced more. A shortage of water can lead to a shortage of energy and vice versa. The two aspects are highly linked and must be considered in a nexus approach for water and energy security as per the sustainable development goals previously discussed. Climate change will also have an effect.

The scale of the water–energy nexus is poised to intensify further in the future with the energy sector projecting a 60% increase in water consumption by 2040 and energy use in the water sector doubling in the same period (OECD/IEA 2016). Any effort to transition to a low-carbon economy, thus, needs to account for the dynamics of the close coupling and interdependencies between these sectors.

Water use at oil and gas operations is not well characterized (DOE 2013). The use of energy in the water sector is also not clearly evaluated (DOE 2014). Energy reliability and resilience is increasingly stressed under declining water resources and climate change. However, if not well managed, efforts to reduce climate change through low-carbon solutions may actually increase water use (OECD/IEA 2016). Approximately 2% of water for irrigation is for biofuel production (WWAP 2009).

Desalination is used to provide drinking water from seawater by removing the salt from the water under a high-pressure membrane system. After use, water is returned to the hydrologic cycle. Distillation is another method of desalting the water. Water is boiled and recondensed from water vapor. Water from desalination is about 3–5 times more expensive than treated freshwater (Pearce 2006). Brine disposal may also be problematic due to the high salt content (Lohan 2008). Engineers are trying to develop combined heat and power methods from (Section 5.4.2) desalination plants as energy requirements are substantial.

5.4.2 Groundwater Management

An effective groundwater management policy must first involve an evaluation of the existing situation including what are the needs of the water, and the laws and regulations to ensure water quality and quantity. As shown in Figure 5.5, the quantities of groundwater must meet the required needs. If this is not the case then time for recharge must be allowed to not deplete this resource. If the quantities are sufficient, then the quality of the groundwater will need to be evaluated to determine for which purpose it is suitable. Industrial or irrigation purposes may not require treatment. As the standards for drinking water quality are more strict; hence, treatment may be required. The treatment is usually performed after pumping. In Chapter 6, evaluation of the most sustainable water treatment processes will be discussed.

As previously mentioned, the 12 principles of green engineering have been developed to improve the sustainability of industrial processes (Anastas and Zimmerman 2003). The second principle indicates that "It is better to prevent waste than to treat or clean up waste after it is formed." Reduction of water use and prevention of contamination of the water should be of highest priority for sustainable water management. This should reduce end of pipe processes, which were the most common solutions.

Therefore, for reducing contaminant introduction, strategies would need to be developed. Aspects of land use such as the ecosystem types, landscapes, and water use can be tracked via groundwater and surface water monitoring

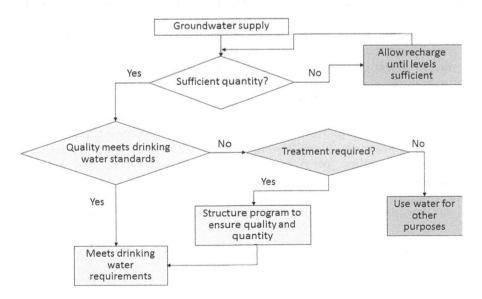

FIGURE 5.5
Flow diagram for management of groundwater management for various purposes. (Adapted from Yong et al. [2014].)

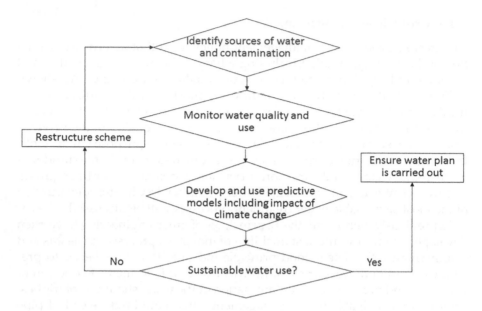

FIGURE 5.6
Development of a program to ensure sustainable water use. (Adapted from Yong et al. [2014].)

for contaminant and nutrient contents and geographical information systems. The information can then be used to optimize water use (Figure 5.6). Education of the society with regard to water usage and its importance is highly important to reduce wastage. Regulations and government policies are also required to protect water quantity.

Resource recovery and stream separation is easier at small scale. Recycling of gray water for toilet flushing is feasible. Industrial wastewater should be treated separately as they can contain various heavy metals and other chemicals. Stormwater is also highly variable and can contain contamination from road runoff. Minimization of water use decreases material and energy requirements. One alternative is reclamation of water for reuse in industry, agriculture, or groundwater recharge. The water will have to respect environmental norms and thus may need to be treated to achieve these norms. Energy can be recovered from wastewater via anaerobic treatment. This produces methane that has similar properties to natural gas.

5.4.3 Case Study on Reclaimed Water Use

Water is necessary for crude oil production and the refining processes. The future development of Canada's oil and gas industry can be threatened due to the limited access to freshwater resources, as the result of climate change and consequent regulatory restrictions. Thus, access to reliable water sources such as reclaimed water can secure future development of the industry.

The search for alternative water resources has increased in response to the freshwater shortage and growing demand for high-quality water.

Reclaimed water is wastewater treated to obtain the quality of freshwater. Assessments showed that reclaimed water use is a sustainable way to meet the growing water demands. The use of reclaimed water not only can meet industry water requirements, but also can considerably reduce freshwater withdrawals by industry. Successful global use of the reclaimed water suggests the feasibility of reclaimed water as an alternative and/or a complementary water supply for Canada's oil activities. Reclaimed water may not only meet the oil industry water quality specifications, but also offers advantages of resilience to climate change, providing additional water supply, supporting industry economic development, and protecting aquatic environments.

One of the key factors in the worldwide acceptance of reclaimed water is that the reclaimed water quality meets the oil industry's chemical and biological water specifications. For example, a study conducted by Qiu et al. (2015) showed that the reclaimed water produced by secondary treatment and ultrafiltration membrane technology at Gold Bar wastewater treatment plant (WWTP) (Edmonton, Canada) can effectively remove nearly 100% of human infectious viruses. Due to the consistent quality of reclaimed water as opposed to the extreme variations in raw river water quality (e.g., seasonal variations in water turbidity), the need for pretreatment of reclaimed water was eliminated and the reclaimed water had a minimum adverse effect on the reverse osmosis (RO) facility at the Suncor refinery (GE Water & Process Technologies 2008b).

The industrial application of reclaimed water showed that the applications of reclaimed water in the oil and gas industry have provided a wide range of direct (monetary) and indirect benefits. Direct benefits are the cost savings resulting from the use of a considerably less expensive reclaimed water than purchasing freshwater which usually involves up-front capital expenditures. Major indirect benefits associated with the use of reclaimed water include (a) increasing water security due to accessibility to an additional good quality water, (b) aquatic environments protection and enhancement, and (c) social effects (ECONorthwest 2012).

One of the major indirect benefits of the reclaimed water use in oil and gas industry is accessibility to a secure and reliable source of high-quality water. This not only can offset the negative effects resulting from water shortage (e.g., due to droughts or other water supply interruptions), but also can support further economic growth by meeting the industry's growing water demands. For example, despite restrictions in extra withdrawals from the North Saskatchewan River, the Suncor Edmonton refinery was able to supply high-quality water for its desulfurization of diesel fuel facility by using reclaimed water (GE Water & Process Technologies 2008a). Similarly, Shell-Canada uses reclaimed water at its Groundbirch natural gas venture to manage the water shortage due to droughts and restrictions on extra withdrawals from the Peace River water (Hamilton 2012; Shell 2012).

Aquatic environmental protection and enhancement is another benefit of reclaimed water use by the oil industry. Aquatic ecosystems need sufficient and good quality water. Due to substantial water withdrawal and contamination, many aquatic ecosystems have been negatively impacted. These ecosystems can be improved if freshwater withdrawal and contamination are reduced. The use of reclaimed water by the oil industry can lead to increased instream flow, especially during the low-flow periods, and water quality improvement (e.g., due to increases in dissolved oxygen), (Miller 2006; Shell 2012). A study conducted by Neufeld (2010) showed that the North Saskatchewan River (Canada) water quality has significantly been improved in recent years due to two main reasons. First, less water withdrawals by the Suncor Edmonton refinery were occurring as it now uses reclaimed water. Second, less nutrients were discharged from the Gold Bar WWTP to the river, since the plant was upgraded to produce low-ammonia and phosphorus reclaimed water.

Social impacts associated with the industrial use of reclaimed water include the ability to avoid water restrictions. This is because reclaimed water is less impacted by climate, and thus it can alleviate the effects of freshwater limitations. Moreover, it can also positively impact the oil industry through developing a healthy industry–community relationship, since the reclaimed water offers an affordable and reliable supply of water to the industry. Therefore, limited freshwater can be used for the community uses, which also lead to conflict reduction (Shell 2012). In summary, the potential application of reclaimed water in Canada's oil industry was investigated. This study highlights the challenges and opportunities of using reclaimed water in the oil industry.

The main driving factors for using the reclaimed water were the local freshwater limitations due to either water shortage or regulatory restrictions, the economic benefits associated with reclaimed water use, and the pressure from other water users (Lahnsteiner et al. 2007; CDM Smith 2012). Reclaimed water quality specifications, availability of financial resources, and technical feasibility were the main factors considered by the oil companies prior to reclaimed water use. Selected technologies for providing suitable reclaimed water for the industry use in most case studies were membranes, mainly RO, due to the high efficiency in removal of concerning contaminants and reliability of the membrane technologies.

The environmental protection, economic benefits, social acceptance, and technical viability were the outcomes of the reclaimed water use by the oil industry. Industry water security was improved because reclaimed water was available continuously and was less affected during drought periods (CDM Smith 2012; Milne 2010). Additionally, reclaimed water use positively affected the environment, since less freshwater was withdrawn by the industry and less wastewater effluent was discharged to the aquatic environment (Beech et al. 2010; CDM Smith 2012). Finally, conflicts over freshwater in water stress regions were reduced because the freshwater was allocated for the domestic purposes (CDM Smith 2012). It is suggested that reclaimed water provides an opportunity for Canada's oil industry to increase their water security for current activities and future expansions. Given the aforementioned benefits and as the further treatment of reclaimed water with the existing technologies is possible, where the high-quality reclaimed water is available, the oil companies should consider it as an additional source of water for their activities.

An oil and gas production and refining company over the years has looked to increase water security including water availability and reliability of supply sources for its facilities, by considering strategies such as potential applications of "process affected water," landfill leachate and wastewater treatment effluent in its facilities. However, the company wanted to examine reliable alternatives such as highly treated municipal wastewater known as reclaimed water. Two options were considered including freshwater from river water (option 1: Business as Usual, BAU) and reclaimed water from a municipal WWTP (Option 2: RW) to assess the potential application of reclaimed water. Using a multi-criteria evaluation tool and identifying various indicators, it was shown that the RW option had an advantage over BAU in the performance of the water management system since it provides an additional water source for the company and is supplemented by freshwater intake through existing infrastructure. The main disadvantage is that there is the potential for scaling and corrosion of existing water treatment plant at the company, depending on the water quality delivered. The RW option improves community well-being, primarily by enabling the local community to allocate additional freshwater for drinking water and other uses within its current freshwater intake system.

5.5 Recycling, Reuse, and Reduce (3 Rs): Aspects of Waste Management

The waste management scheme in terms of preferences follows prevention of waste generation, waste minimization, then reuse, recycling, and the least favorable, landfilling (Figure 5.7). There are many challenges in waste management. Incorporating the end of life in the design process can reduce the difficulties as mentioned previously.

Solid waste disposal in landfills requires large amounts of space that is less and less available. Emissions of carbon dioxide, methane, and leachates are occurring. Valuable resources can be obtained from the waste for energy or materials. Lack of technologies for dealing with mixed materials and a lack of suitable markets can make recycling difficult. Energy recovery through incineration is an option. There has been a lot of mistrust from the public regarding the safety of incineration, however. Prevention and reduced

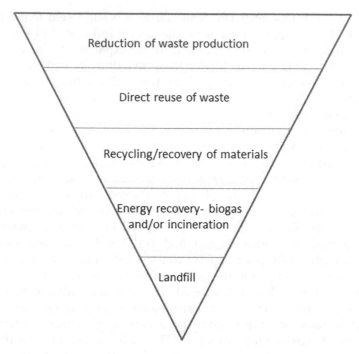

FIGURE 5.7
Hierarchy of waste management options with the most preferable option at the top. Waste management, however, should be an integrated approach.

production of waste are preferred. Some options can be seen in Figure 5.8 (Azapagic 2011).

Composting is an option that has gained a lot of interest recently for reducing the input of organics into landfills. Food and garden wastes can be used to produce a soil improver. However, this option can reduce the production of biogas in landfill with biogas recovery for energy production. Anaerobic digestion to produce energy in the form of biogas is another option. A proposed life cycle is shown in Figure 5.8.

In the EU, practices vary but there is an overall trend to reducing landfill disposal. Sweden has banned landfilled entirely. In the United States, the EPA has adopted a similar strategy to the EU, reduction of waste sources, then recycling or composting followed by disposal with energy recovery and landfilling. In Asia, countries such as Japan have extensive recycling programs and wastes for incineration are collected in transparent bags to ensure there are no hazardous materials such as batteries.

Incineration includes waste handling, the incinerator and boiler, energy recovery and generation, air pollution control equipment, and ash handling. Temperatures of 980°C–1,090°C are used. Heat is recovered in the form of steam or hot water. Combined heating and power systems (CHPs) are the most efficient as 85% of the energy is recovered compared to electricity

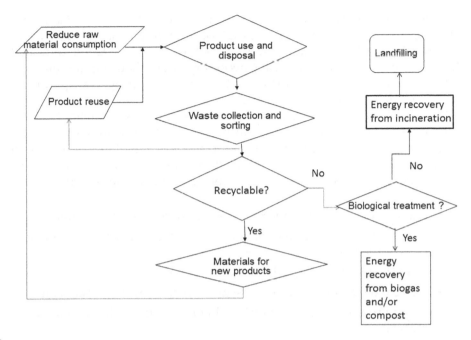

FIGURE 5.8
Integrated life cycle of waste management.

generation which recovers 35% only. Designing incineration depends on waste composition and quantity, and requires specification of the incinerator type (moving grate, fluidized bed, or rotary kiln) and the energy recovery public acceptance can be a major issue for building a new plant despite technical, economic, and environmental feasibility. Measures to prevent air emissions of nitrogen, sulfur and carbon oxides, HF and HCl, particulates and heavy metals and dioxins and furans are necessary.

Reduction of wasted materials must be considered with energy use. Levis et al. (2014) found that diversion was not the most energy-efficient option. Although the concept of circular economy has become popular in Europe, it is still very far from this. In 2005, Haas et al. (2015) indicated that out of 58 Gt/year of materials extracted, there is an output of 41 Gt/year that is not recycled. Rating systems (Envision, SITES, and INVEST) are now employed to reduce waste. Credits are given for recycling of materials, reusing wastes, and reducing energy and water use.

Construction and demolition (C&D) waste is the waste from the construction, operation, and demolition of infrastructure. Materials such as rare-earth elements and aggregates for construction are becoming more and more scarce. Therefore, recovery of these materials is very important. Townsend et al. (2014) indicated that about 70% of the waste is recycled. This can be broken down as 99% of RAP, 85% of bulk aggregates, and 35% of mixed C&D

wastes. Asphalt and concrete are the most fractions of C&D waste. Gambatse and Rajendran (2005) compared the LCA of reinforced concrete and asphalt pavements. Asphalt generated less overall waste. Field (2010) indicated that a life cycle approach at least is necessary when choosing the construction material. This includes the extraction, manufacturing, transportation, construction, use and demolition steps.

Waste prevention, minimization, or reduction is the most favorable. Not only should the amount of the waste be reduced, but also the toxicity. As previously discussed, methods for this are design for increased durability and reduced maintenance, and designing for disassembly and deconstruction. Calkins (2012) has suggested several approaches for designing for deconstruction:

- Site selection and adaptation for maximum flexibility over time
- Ensure all materials and methods are documented to facilitate deconstruction and disassembly
- Select materials with good reuse or recycling potential
- Design connections that are accessible and document for disassembly
- Avoid procedures such as finishes that reduce the ability to reuse or recycle
- Soil use.

Buildings are the most common structure that can be reused. Basu et al. (2015) had indicated that foundations can be cost effective and are becoming more common. Road material and retaining walls can also be reused.

Reprocessing or remanufacturing of materials is the most common approaches for recycling. Calkins (2012) indicated that aggregates, asphalt, bricks, concrete, metals, plastic, soil materials, and wood. Fly ash, industrial by-product can be used to replace cement. Leaching of metals and other toxic components from the products must be evaluated before use. Waste glass can also be crushed, screened, and added in a road subbase to reduce the use of aggregates. Roofing shingles can be used as base and subbase materials (Basu and Puppala 2015).

The Building Environmentally and Economically Sustainable Transportation-Infrastructure-Highways (BE^{2}ST-in-Highways) program has been used to evaluate the sustainability of four pavement designs in Baraboo, Wisconsin (Lee et al. 2013). Recycling of materials (recycling of asphalt with or without fly ash) reduced the GWP and energy consumption. The properties of the recycled materials were excellent and enabled less material to be used and less maintenance was required.

Waste to energy, zero waste and upcycling approaches are all viewed as sustainable practices. Some indicators (Wright 2017) for wastes are amount of waste diverted from landfills, waste in water effluents, energy required to transport waste to landfills, recovery percentage of electronic waste.

5.5.1 Case Study of the Saint Michel Environmental Complex

The Saint Michel Environmental Complex of the City of Montreal, Canada occupies 192 hectares (Ville de Montréal, revalorisation du CESM http://ville.montreal.qc.ca/portal/page?_pageid=7237,75372019&_dad=portal&_schema=PORTAL). In the past, it was the Miron Quarry that was converted in 1968 to a landfill of 75 hectares. Today it is the second largest green space in Montreal. The Complex now consists of a sorting center for recyclable materials, the production of electricity from biogas produced at the site and composting and wood chipping facilities. The area is being converted into a large green space by 2023. Currently, there is a park around the outer limits of the park that include 5 km of bike paths and areas for several games. The environmental challenges at the site include recuperation of the biogas, treatment of the leachate, environmental monitoring, and the construction of the Frédéric-Back Park. To complete the construction of the park, excavated soil of residential or almost residential quality is being used.

Since 1996, Gazmont undertook the challenge of converting the biogas produced from the waste to electricity. At that time, the gas was burned by flares producing carbon dioxide emissions. Approximately, 375 wells for capturing the gas are placed at a depth of 30 m. The wells with the assistance of compressors bring the gas to the surface that is transported to the Gazmont facility. In 2009, 10,500 m^3 of biogas was extracted per hour. The landfill is now closed. However, 23 MW are still produced that is fed to the grid (equivalent to the needs of 15,000 residences). The flares are now only used when the facility is taken offline for maintenance.

The leachate from the landfill, that is 80 m in depth at the bottom of the quarry, is recovered by a pumping system at a rate of 1,400 m^3/day. Air is injected for treatment to reduce pollutants such as sulfur. The water is then sent to the sewer system for treatment at the wastewater treatment center of Rivière-des-Prairies.

In the fall of 2009, the City of Montreal undertook to cover 30 out of the 75 hectares of the landfill, which will be the site of the future park. Although the biogas is captured by the system, some still escapes. The final cover was placed to inhibit the escape of the gas while providing a base for plant growth.

Since 1984, Montreal established a dialogue with the citizens such as offering guided visits to the site. Since 1989 approximately 135,000 had visited the site. Since 2004, TOHU established a welcome center and assumed the responsibility for the guided visits of the public. Since 2009, the Groupe TIRU performs the collection and sorting of recyclable materials. No sorting is done by the citizens. All the recyclables are placed in a collection bin that is collected at roadside. At the sorting center, the materials are subjected to automated or manual sorting. The trucks arrive with the materials and are weighed.

The first step of sorting removes large objects and other nonrecyclables and open and removes plastic bags. Next, cardboard, newspapers, paper, glass, plastics, and metals are sorted by mechanical separators followed by manual sorting of the papers, separating cardboard, newspaper, and mixed paper according to the needs of the recyclers.

Next, the glass, plastic, and metals are sorted mechanically. Glass is not separated by color. Ferrous metals are separated by means of a magnet. The plastics are separated optically. Sorting is adjusted according to the market. Nonferrous metals such as aluminum are recovered. Glass, plastic, and metals are sent to another recycling center. Daily, the sorting center receives 150 trucks and about 800 metric tons of materials are sent to recyclers.

Since 1989, the Complex composts thousands of metric tons of dead leaves collected from the Montreal area in bulk or in bags. The leaves are deposited in windrows and are turned frequently until the winter. The turning then restarts in the spring. The compost is used for the final cover of the landfill and is given to the citizens freely twice a year.

5.6 Conclusions

In this chapter, practices to reduce use of materials, energy and water, and waste production were examined. With regard to materials, renewable materials should be used as much as possible due to limited resources. Materials in the production process and final product should be biodegradable and nontoxic to avoid accumulation in the environment and impact on the environment and human health. Recycling or reuse of materials should be employed to reduce consumption of raw resources. Design for reuse and disassembly practices are useful for extending product life and usefulness. Material selection and design procedures are thus highly important.

Engineers are particularly engaged in the choice of the energy source and improving energy efficiency. GHG emissions and other toxic emissions are particularly tied to energy use and thus energy efficiency is essential to mitigate climate change and environmental damage. Renewable energies will conserve resources. Development of new sources of energy is particularly important.

Water quality and quantity are also highly important issues for engineers, from the water infrastructure aspect, the extensive use in industry and its relation with energy production. Its use needs to be balanced with other uses in the community. Water reuse and recycling need to be increased. Gray water can be employed for many purposes other than drinking water. Water use is substantial in energy production and energy is required for water treatment. Therefore, the nexus of water and energy is a field of importance for future research.

To reduce waste, the most appropriate strategy is to minimize waste generation. This can be done through sustainable design procedures. Minimization of waste during production processes, selection of materials for reusability, and design of products for reuse are some practices to achieve this.

An integrated life cycle approach for product and process design is essential. Sustainability assessments can assist in the selection of materials, energy, water use, and waste minimization through the use of indicators, as will be discussed in Chapter 7.

Acknowledgment

The author would like to acknowledge the contribution of Dr. Nayereh Saborimanesh to the case study on water reclamation and the funding support of Mitacs and Husky Energy for the project and the technical support of many others.

References

Abbasi-Garravand, E., C.N. Mulligan, C.B. Laflamme and G. Clairet, 2016. Role of two different pretreatment methods in osmotic power (salinity gradient energy) generation. *Renewable Energy*, 96: 98–119.

Abbasi-Garravand, E., C.N. Mulligan, C.B. Laflamme, G. Clairet, 2017. Identification of the type of foulants and investigation on the membrane cleaning methods for PRO processes in osmotic power application. *Desalination*, 421(1): 135–148.

Achilli, A. and A.E. Childress, 2010. Pressure retarded osmosis: From the vision of Sidney Loeb to the first experimental installation—Review. *Desalination*, 261: 205–211.

Allenby, B.R., 2012. *The Theory and Practice of Sustainable Engineering*, Upper Saddle River, NJ: Prentice Hall.

Anastas, P. and J. Zimmerman, 2003. Design through the 12 principles of green engineering. *Environmental Science and Technology*, 37: 94A–101A.

APA (American Planning Association), 2016. Policy Guide on Water. www.planning. org/policy/guides/adopted/water/. Accessed on June 6, 2018.

ASTM, 2010. Basic Assessment and Management of Greenhouse Gases, E2725, December 10, 2019.

Azapagic, A., 2011. Municipal solid waste management recovery energy from waste. In: *Sustainable Development in Practice: Case Studies for Engineers and Scientists*, 2nd edition, A. Azapagic and S. Perdan (eds), Chichester: John Wiley & Sons, pp. 261–325,

Bartos, P.J., 2002. SX-EW copper and the technology cycle. *Resources Policy*, 28: 85–94.

Basu, D. and A.J. Puppala, 2015. Principles of sustainability and their applications in geotechnical engineering. In: *15th Pan American Conference on Soil Mechanics and Geotechnical Engineering*, Buenos Aires, Argentina/Clifton, VA: IOS Press, (Keynote Lecture).

Basu, D., A. Misra and A.J. Puppala, 2015. Sustainability and geotechnical perspectives and review. *Canadian Geotechnical Journal*, 52(1): 96–113.

Beech, I.B., J.A. Sunner and K. Hiraoka, 2010. Microbe-surface interactions in biofouling and biocorrosion processes. *International Microbiology*, 8(3): 157–168.

Butcher, A.P., J.J.M. Powell and H.D. Skinner (eds), 2006. *Whole Life Cost and Environment Impact Case Studies. Reuse of Foundations for Urban Sites: a Best Practice handbook*, Berkshire: IHS BRE Press, pp. 116–119.

Calas, G., 2017. Mineral resources and sustainable development. *Elements: An International Magazine of Mineralogy, Geology and Petrology*, 13(5): 301–306.

Calkins, M., 2012. Site design: Materials and resources. In: *The Sustainable Sites Handbook. A Complete Guide to the Principles, Strategies and Practices for Sustainable Landscapes*, M. Calkins (ed), Hoboken, NJ: Wiley, pp. 323–428.

Chamley, H., 2003. *Geosciences, Environment and Management*, Amsterdam: Elsevier, p. 450.

Cheng, V.K.M. and G.P. Hammond, 2017. Life-cycle energy densities and land take requirements of various power generators: A UK perspective. *Journal of Energy Institute*, 90: 201–213.

Daigger, G.T. and G.V. Crawford, 2005. Incorporation of biological nutrient removal (BNR) into membrane bioreactors (MBRs). In: *Proceedings of the IWA Specialized Conference, Nutrient Management in Wastewater Treatment Processes and Recycle Streams*, Krakow, Poland, September 19–21, 2005.

De Villiers, J.P.R., 2017. How to sustain mineral resources: Beneficiation and mineral engineering opportunities. *Elements: An International Magazine of Mineralogy, Geology and Petrology*, 13(5): 307–312.

DOE (U.S. Department of Energy), 2013. Effects of Climate Change on Federal Hydropower: Report to Congress, Washington, DC. www.energy.gov/eere/water/downloads/effects-climate-change-federal-hydropower-report-congress. Accessed November 12, 2018.

DOE (U.S. Department of Energy), 2014. The Water-Energy Nexus: Challenges and Opportunities, Washington, DC. www.energy.gov/downloads/water-energy-nexus-challenges-and-opportunities. Accessed November 12, 2018.

ECONorthwest, 2012. *King County Reclaimed Water Comprehensive Plan Benefit-Cost Analysis of Reclaimed Water Strategies*, King County: Department of Natural Resources and Parks, Wastewater Division.

European Commission, 2018. End of Life Vehicles. http://ec.europa.eu/environment/waste/elv/index.htm. Accessed June 6, 2018.

Field, R., 2010. *Materials Introduction. Sustainability Guidelines for the Structural Engineer.* D.M. Kestner, J. Goupil and E. Lorenz (eds), Reston, VA: ASCE.

Fukue, M., S. Ono and Y. Sato, 2011. Cementation of sands due to microbiologically-induced carbonate precipitation. *Soils and Foundations*, 51(1): 83–93.

Fukue, M., S. Ono, Y. Sato, I. Sakamoto, T. Iwata and C.N. Mulligan, 2013. Microbial cementation of dry sands by injecting microbes and chemical agents. In: *Proceedings of the 3rd Annual International Conference on Advances in Biotechnology, BIOTECH 2013*, Singapore, pp. 17–22.

Gagnon, B., R. Leduc and L. Savard, 2012. From a conventional to a sustainable engineering design process: Different shades of sustainability. *Journal of Engineering Design*, 23(1): 49–74.

Gambatse, J.A. and S. Rajendran, 2005. Sustainable roadway construction: Energy consumption and material waste generation of roadways. In: *Proceedings of Construction Research Congress 2005*, Broadening Perspectives, Reston, VA: ASCE, pp. 104–110.

GE Water & Process Technologies, 2008a. City of Edmonton Gold Bar Wastewater Treatment Plant. GE Water & Process Technologies. p. 2.

GE Water & Process Technologies, 2008b. Petro-Canada's Largest Oil Refinery Looks to GE For Municipal Effluent Water Reuse, GE Water & Process Technologies. p. 4.

GEMI, 2015. *Connecting the Drops toward Creative Water Strategies: A Water Sustainability Tool.* Gandhinagar: GEMI. http://gemi.org/. Accessed July 15, 2018.

Haas, W., F. Krausmann, D. Wiedenhofer and M. Heinz, 2015. How circular is the global economy? An assessment of material flows, waste production and recycling in the European Union and the world in 2005. *Journal of Industrial Ecology*, 19(5): 765–777.

Hamilton, G., 2012. Shell uses recycled water for Dawson Creek fracking, in Vancouver Sun. 09/08/2012.

Heinberg, R., 2011. *The End of Growth: Adapting to our New Economic Reality*, Gabriola Island, BC: New Society Publishers.

Hoekstra, A.Y. and A.K. Chapagain, 2007. Water footprints of nations: Water use by people as a function of their consumption pattern. *Water Resources Management*, 21: 35–48.

Jones, R. and N. Dixon, 2011. Sustainable development using geosynthetics: European perspectives. *Geosynthetics*, 29(2). http://geosyntheticsmagazine.com.

Kabbes, K., J. Reichenberger, C. Briggs, C. Davidson and A. Perks, 2017. Water resources: sustaining quality and quantity. In: *Engineering for Sustainable Communities: Principles and Practices*, W.E. Kelly, B. Luke and R.N. Wright (eds), Roston, VA: ASCE Press, pp. 237–254.

Kim, Y.C. and M. Elimelech, 2013. Potential of osmotic power generation by pressure retarded osmosis using seawater as feed solution: Analysis and experiments. *Journal of Membrane Science*, 429: 330–337.

König, K.W., 1999. Rainwater in cities: A note on ecology and practice. In: *Cities and the Environment: New Approaches for Eco-Societies*, T. Inoguchi, E. Newman and G. Paoletto (eds), Tokyo: United Nations University Press, pp. 203–215.

Lahnsteiner, J., F. Klegraf, R. Mittal and P. Andrade, 2007. Reclamation of wastewater for industrial purposes–advanced treatment of secondary effluents for reuse as boiler and cooling make-up water. In: *Proceedings of the 6th Specialist Conference on Wastewater Reclamation and Reuse for Sustainability*, Antwerp, Belgium, October 9–12, 2007.

Landsber, D.R. and M.R. Lord, 2009. *Energy Efficiency Guide for Existing Commercial Buildings. The Business Case for Building Owners and Managers*, Atlanta: ASHRAE.

Lee, J., T.B. Edil, C.H. Benson and J.M. Tinjuan, 2013. Building environmentally and economically sustainable transportation infrastructure. Green highway rating system. *Journal of Construction Engineering and Management*, 139(12): A4013006.

Levis, J.W., M.A. Barlaz, J.F. DeCarolisand, S.R. Ranjithan, 2014, Systematic exploration of efficient strategies to manage solid waste in U.S. municipalities: Perspectives from the solid waste optimization life-cycle framework (SWOLF). *Environmental Science and Technology*, 48(7): 3525–3631.

Lohan, T. (ed), 2008. *Water Consciousness*, San Francisco, CA: AlterNet Books.

Mata, T.M., A.A. Martins, S.K. Sikdar and C.A.V. Costa, 2011. Sustainability considerations of biodiesel based on supply chair analysis. *Clean Technology and Environmental Policy*, 13(5): 655–671.

Maupin, M.A., J.F. Kenny, S.S. Hutson, J.K. Lovelace, N.L. Barber and K.S. Linsey, 2014. *Estimated Use of Water in the United States in 2010*. Reston, VA: U.S. Geological Survey Circular 1405.

Miller, G.W., 2006. Integrated concepts in water reuse: Managing global water needs. *Desalination*, 187(1): 65–75.

Milne, N., 2010. *Guidance for the Use of Recycled Water by Industry*. Institute for Sustainability and Innovation, CSIRO Land and Water, Victoria University, p. 182. www.vu.edu.au/institute-for-sustainability-and-innovation-isi/publications.

Neufeld, S., 2010. North Saskatchewan River Water Quality. The Edmonton Sustainability Papers-Discussion Paper 4, p. 14. www.epcor.com/products-services/water/Documents/north-saskatchewan-river-water-quality.pdf. Accessed November 12, 2018.

NOAA (National Oceanic and Atmospheric Administration), 2012. Incorporating Sea Level Rise Scenario at the Local Level. www.ngs.noaa.gov/PUBS_LIB/SLCScenariosLL.pdf.

OECD/IEA, 2016. Water-Energy Nexus, Excerpt from the World Energy Outlook 2016. Paris, France. www.iea.org/publications/freepublications/publication/WorldEnergyOutlook2016ExcerptWaterEnergyNexus.pdf. Accessed November 12, 2018.

Pearce, F., 2006. *When the Rivers Run Dry*, Boston, MA: Beacon Press.

Puppala, A., J.T. Das, T.V. Bheemasetti and S.S. Congress, 2018. *Sustainability & Resilience in Transportation Infrastructure Geotechnics*, GEOSTRATA, pp. 42–48.

Qiu, Y., B.E. Lee, N. Neumann, N. Ashbolt, S. Craik, R. Maal-Bared and X. Pang, 2015. Assessment of human virus removal during municipal wastewater treatment in Edmonton, Canada. *Journal of Applied Microbiology*, 119(6): 1729–1739.

Randolph, J. and G.M. Masters, 2008. *Energy for Sustainability: Technology, Planning, Policy*, Washington, DC: Island Press.

Robertson, M., 2017. *Sustainability Principles and Practice*, London and New York: Routledge.

Rubin, E.S., 2001. *Introduction to Engineering and the Environment*, New York: McGraw Hill.

Saborimanesh N. and C.N. Mulligan, 2015. Effect of sophorolipid biosurfactant on oil biodegradation by the natural oil-degrading bacteria on the weathered biodiesel, diesel and light crude oil. *Journal of Bioremediation and Biodegradation*, 6: 314.

Saskpower, 2018. Boundary Dam Carbon Capture Project. www.saskpower.com/our-power-future/infrastructure-pfrojects/carbon-capture-and-storage/boundary-dam-carbon-capture-project. Accessed June 1, 2018.

She, Q., Y.K.W. Wong, S. Zhao and C.Y. Tang, 2013. Organic fouling in pressure retarded osmosis: Experiments, mechanisms and implications. *Journal of Membrane Science*, 428: 181–189.

Shell, 2012. Shell and Dawson Creek to Conserve Water in British Columbia. Shell Canada. www.shell.ca/en/aboutshell/media-centre/news-and-media-releases/2012/0907dawson-creek.html.

Sikdar, S.K., D. Sengupta and R. Mukherjee, 2017. *Measuring Progress towards Sustainability: A Treatise for Engineers*, Cham, Switzerland: Springer International Publishing.

Smith, C.D.M, 2012. Chapter 9: Global Experiences in Water Reuse; USEPA Guidelines for Water Reuse. EPA/600/R-12/618, US Environmental Protection Agency, p. 643. http://nepis.epa.gov/Adobe/PDF/P100FS7K.pdf. Accessed July 15, 2018.

Townsend, T.G., C. Wilson and B. Beck, 2014. The Benefits of Construction and Demolition Materials Recycling in the United States. A CDRA white paper. www.usagypsum.com/wp-content/uploads/2016/05/CDRA-White-paper-executive-summary.pdf. Accessed November 12, 2018.

UNEP, 2011. Recycling Rates of Metals. A Status Report. UNEP. www.resourcepanel.org/file/381/download?token=he_rldvr. Accessed November 12, 2018.

USBR (U.S. Bureau of Reclamation), 2016. West-Wide Climate Risk Assessments. Hydroclimate Projections. www.usbr.gov/climate/secure/docs/2016secure/wwcra-hydroclimateprojections.pdf.

Weeks, J., 2005. Building an energy economy on biodiesel. *Biocycle*, 46(7): 67–68.

Wright, R., 2017. Sustainable land use. In: *Engineering for Sustainable Communities: Principles and Practices*, W.E. Kelly, B. Luke and R.N. Wright (eds), Roston, VA: ASCE Press, pp. 157–178.

WWAP (United Nations World Water Assessment Programme), 2009. The United Nations World Water Development Report 3. Paris: Water in a Changing World, United Nations Educational, Scientific and Cultural Organization (UNESCO).

Yong, R.N., C.N. Mulligan and M. Fukue, 2014. *Sustainable Practices in Geoenvironmental Engineering*, 2nd edition, Boca Raton, FL: CRC Press.

Yong, R.N., M. Nakano and R. Pusch, 2010. *Containment of High Level Radioactive and Hazardous Solid Wastes with Clay Barriers*, London: Spon Press, Taylor and Francis, 468 pages.

6

Management of Contaminants in the Environment

6.1 Introduction

Pollutants can be emitted into the atmosphere, water, and soil environment as indicated in Chapter 1. They can accumulate in organisms or be transported or transformed in the environment. A number of mechanisms are shown in Figure 6.1. Some examples of contaminants include dry cleaning

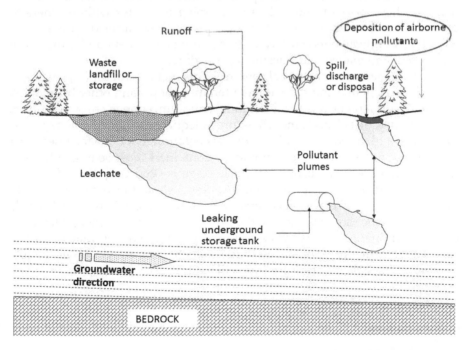

FIGURE 6.1
Schematic representation showing pollutant plumes from spills, a leaking underground storage tank or landfill, runoff, wastewater discharge, and airborne pollutant deposition. (Adapted from Yong and Mulligan [2004].)

solvents, various oils including lubricating oil, automotive oil, hydraulic oil, fuel oil, and biosolids from wastewater plants, processing wastes from various industries such as pulp and paper deinking sludges and organic and inorganic aqueous wastes and wastewaters. There are several ways in which wastes and contaminants can be classified, such as:

- The level of toxicity
- As inorganic or organic substances
- The type of industry, such as pulp and paper, petroleum, mining, forest, electronic, pharmaceutical, etc.
- The class of chemicals such as pesticides, solvents, etc.
- The nature of impact or threat such as biological physical, chemical, etc.
- The type of receptor (land, water, human, biota, etc.)

Soil contamination is the result of accidental spills and leaks, generation of chemical waste leachates and sludges from cleaning of equipment, residues left in used containers and outdated materials. Inadequate methods for waste storage, treatment, and disposal have led to the contamination of many sites, particularly at bankrupt and abandoned manufacturing plants. Some of these disposal methods are dumping of wastes without liners to protect the soil and groundwater, inappropriate siting of the dumping areas close to groundwater and surface waters and pierced liners under the waste disposal due to inadequate placement procedures.

Smaller generators of chemical contaminants include improperly managed landfills, automobile and railway maintenance shops, and other businesses. The more common heavy metals include cadmium, copper, chromium, iron, lead, mercury, nickel, and zinc. Therefore, adequate methods for storage and disposal are required for avoidance of the contamination. The facilities must be designed to contain the contaminants. Containers must be in good condition and monitored to ensure that no leakage is occurring. Procedures for disposal must be correctly followed. Processes should also be designed to minimize waste and emission generation as indicated in Chapter 5. Materials that are nontoxic should be used in the process as much as possible. However, once contamination occurs in the soil or water, remediation may be necessary. The remediation or treatment thus should be performed in a sustainable manner as possible. Practices for this will be discussed in this chapter.

6.2 Soil Remediation

Two options are available for disposal of contaminated soil: (a) disposal in a secure landfill or disposal facility and (b) treatment of the contaminated soil and reuse of the treated soil. Option (a) is not a preferred option as it is not sustainable. Treatment of contaminated soil can be an expensive procedure, especially when the quantities are large.

However, once the contamination takes place, a variety of *in situ* and *ex situ* remediation techniques exists. To evaluate the most appropriate and sustainable technology, the procedure in Figure 6.2 should be followed. In general, technologies are classified as physical, chemical, and biological. They can be either *in situ*, meaning that the treatment is at the site (reducing transportation requirements) or *ex situ* where the soil is excavated and treated. *Ex situ* techniques include contaminant solidification/stabilization, incineration, vitrification, physical separation, washing, and biological treatment processes. *In situ* processes include (a) bioremediation, (b) extraction methods for air or steam stripping or thermal treatment for volatile compounds or flushing for soluble components, (c) chemical treatments for oxidation and (d) stabilization/solidification with cements, lime, and other additives for heavy metal or organic contaminants. Phytoremediation although less developed has also been used. The most suitable types of plants must be selected based on pollutant type and recovery techniques for disposal of the contaminated plants. Other technologies related to

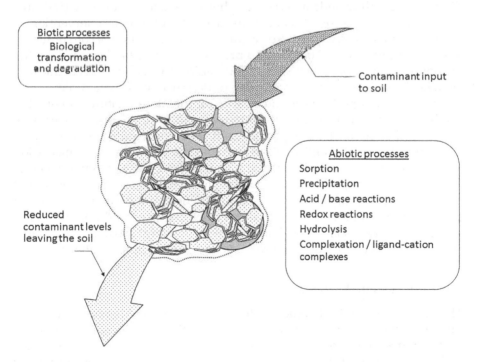

FIGURE 6.2
Natural attenuation mechanisms of pollutants in soil. (Adapted from Yong and Mulligan [2004].)

nanotechnologies are also being developed (Babaee et al. 2018). More detail on the various technologies can be found in Yong et al. (2014).

6.2.1 Natural Attenuation

The various naturally occurring processes responsible for the reduction of the concentration and toxicity of contaminants in the soil are known as natural attenuation (Figure 6.2). The processes include physical, chemical and biologically mediated mass transfer, and biological transformation. The American Society for Testing and Materials (ASTM 1998) defines *natural attenuation* as the "reduction in mass or concentration of a compound in groundwater over time or distance from the source of constituents of concern due to naturally occurring physical, chemical, and biological processes, such as; biodegradation, dispersion, dilution, adsorption, and volatilization." The U.S. Environmental Protection Agency (USEPA 1999c) uses the term monitored natural attenuation and defines it as: "the reliance on natural attenuation processes (within the context of a carefully controlled and monitored site cleanup approach) to achieve site-specific remediation objectives within a time frame that is reasonable compared to that offered by other more active methods. The 'natural attenuation processes' that are at work in such a remediation approach include a variety of physical, chemical, or biological processes that, under favorable conditions, act without human intervention to reduce the mass, toxicity, mobility, volume, or concentration of contaminants in soil or groundwater. These in-situ processes include biodegradation; dispersion; dilution; sorption; volatilization; radioactive decay; and chemical or biological stabilization, transformation, or destruction of contaminants."

Guidelines and protocols for application of monitored natural attenuation (MNA) as a treatment procedure in remediation of contaminated sites have been issued. Adaption for site specificities, however, is required. A general protocol, from Yong and Mulligan (2004) for considering MNA as a remediation tool is shown in Figure 6.3. Due to the sustainability benefits of natural attenuation, it can be integrated into a remediation scheme as either a pretreatment or a finishing step after the remediation. Reduction and elimination of the presence of contaminants in the soil is essential. Engineering the natural attenuation capability of soils through the use of geochemical, biological, and nutrient enhancement will provide more sustainable management options.

6.2.2 Sustainable Remediation

The ASTM has devised some guidelines for site remediation to incorporate sustainability objectives and to reduce the environmental footprint of the remediation process (ASTM 2013, 2016). To incorporate sustainability objectives, all three aspects (environmental, social, and economic) must be considered. Various elements and best management practices (BMPs) and the processes for implementation are indicated in Figure 6.4.

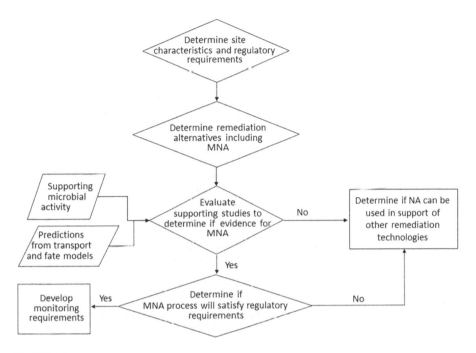

FIGURE 6.3
General protocol for selecting MNA for remediation. (Adapted from Yong and Mulligan [2004].)

Sustainability objectives are the integration of sustainability into the project. The core elements including those of the USEPA (2009) are:

- Reduction of energy use and maximization of renewable energy (suggestions include using energy-efficient equipment using renewable energies)
- Minimization of emissions of air pollutants and greenhouse gases (GHGs, suggestions include reducing dust and contaminant transport, using equipment that is efficient or with clean fuels)
- Minimization of the use of water and the impact on water resources (suggestions include conservation of water, reclamation of water and reuse, using BMPs for erosion, stormwater, or sedimentation control, revegetation with plants that are water efficient)
- Reduction, reuse, and recycling of waste and materials (suggestions include use of renewable, local or recycled materials, reusing materials such as fly ash in concrete, recycling of construction and demolition waste)
- Protection of land and ecosystems (suggestions include minimization of land disturbance and destruction of natural environments and noise and light)

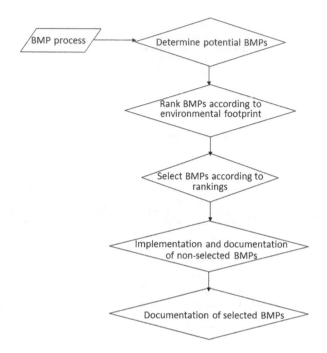

FIGURE 6.4
BMP process by ASTM. (Adapted from ASTM [2013].)

A quantitative analysis can be undertaken via a footprint analysis or life cycle assessment (LCA). ISO 14044 (2006) or EPA documents (USEPA Life Cycle Assessment, Principles and Practice, EPA/600/R-06/060 (May 2006) or USEPA Methodology for Understanding and Reducing a Project's Environmental Footprint, EPA 540-R-12-002 (February 2012b) can be used for these procedures.

The latter includes meetings or hearings, website information, media announcements, and other publicly available information. A detailed list of BMPs is shown in the appendix of the guideline. The BMP process should be included to reduce the environmental footprint and document the processes undertaken.

An ASTM guide for integration of sustainable objectives into Cleanup (E2876-13) has been developed, which includes the environmental, social, and economic aspects. The various aspects of a project are shown in Figure 6.5. The first phase involves planning and scoping of the project. This is followed by information gathering, including stabling the sustainability objectives. Data needs and activities are identified for the project. The sustainable core elements include air emissions, community involvement, economic impacts on the local community and government, efficient and economic cleanup, and minimization of energy use. Other elements include enhancing the human environment, reduction of land and ecosystem impacts, vitality of the

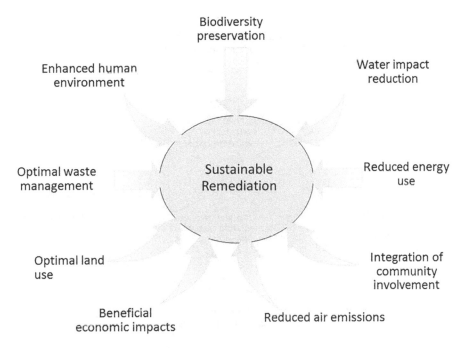

FIGURE 6.5
Elements to be incorporated into a sustainable remediation. (Adapted from ASTM [2013].)

local community, minimization of materials and wastes, water impact, and inclusion of stakeholders.

The ASTM guidelines are mainly focused on the United States due to regulatory aspects but the process can be applied in other locations. A suggested process is indicated in Figure 6.6. BMPs should be adopted to reduce the environmental footprint of the remediation. The steps include the site assessment, evaluation of the technologies for remediation, carrying out the remediation, and monitoring and optimization of the process. Some elements to reduce the footprint are summarized in Table 6.1. A more extensive greener cleanup BMP table for various technologies is included in the guideline. For example, for pump and treatment, bio-based products can be used such as biological surfactants. Mulligan (2014) has shown that biodegradable, nontoxic products called biosurfactants (e.g., rhamnolipids and sophorolipids) can be produced from waste materials and employed for soil flushing or washing of metal and organic contaminants or for enhanced biodegradation of organic pollutants. Biosurfactant applications for remediation of contaminated soil and water are promising due to their biodegradability, low toxicity, and critical micelle concentration and high effectiveness in enhancing biodegradation and affinity for metals. Biosurfactants can remove heavy metals through the mechanisms of solubilization, complexation, and ion exchange. Most research has involved rhamnolipids. Other biosurfactants and process scale-up need further investigation.

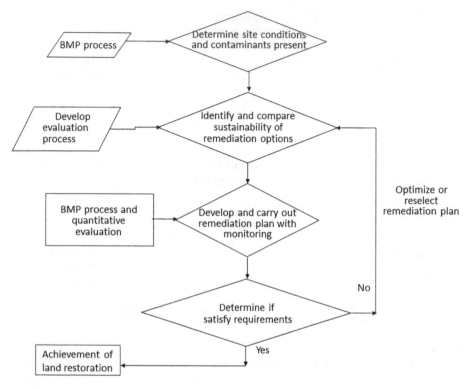

FIGURE 6.6
Greener remediation scheme by ASTM. (Adapted from ASTM [2013].)

TABLE 6.1

Elements to Reduce Environmental Footprint of Remediation (ASTM 2016)

Objective	Means of Achieving Objective
Energy reduction and use of renewable energy	Employment of energy-efficient equipment, use of wind or solar energy
Reduction of air and GHG emissions	Minimize dust, use of equipment with emission controls, use of clean fuels or hybrids
Reduction of water use and impact on resources	Use of water-efficient equipment, water reclamation for reuse, use of green technologies for erosion and runoff control
3Rs of material use	Use of concrete from coal combustion products, use of renewable or certified products such as wood, use of locally generated materials
Protection of land and ecosystems	Minimization of natural vegetation disturbance, destruction of natural habitats, restoration of biodiversity

Other components to be incorporated in sustainable remediation can include:

- Using recycled concrete for pipe bedding, landscaping
- Using natural gas, low emission and noise or clean diesel generators or renewable energy for equipment
- Employing gravity flow where possible
- Recharging groundwater with uncontaminated groundwater
- Reducing waste materials with reusable equipment
- Using on-site analysis to avoid the need for off-site shipping
- Using native plants to restore biodiversity
- Minimizing tree and vegetation removal and traffic routes to reduce site disturbance
- Using pervious materials for pavements
- Collecting rainwater for dust control and other uses

Tools such as LCA and footprint analysis enable documentation of the wastes and emissions produced once the remediation components have been identified. Emissions, energy, and water can all be tracked in the footprint analysis. LCA tools have been previously discussed in Chapter 3.

6.2.3 Brownfield Redevelopment

There are numerous benefits of restoring contaminated abandoned sites named brownfields. Some of these include reducing urban sprawl, increased tax revenue, improving land, public health and safety by improving air and water quality, and reducing GHG emissions (NRTEE 2004). Preservation of 4.5 ha of greenland can be done for every hectare of remediated brownfield. Land use is more compact, thus increasing efficiencies.

For redevelopment of urban-contaminated sites, various parameters need to be determined. Some of these factors include site characterization of the soil (mineral, texture, geochemical characteristics, groundwater flow) and contaminants present, the factors influencing fate and mobility of the contaminants such as dilution, sorption/desorption, biodegradation and transformation, and other factors such as climate, and microbial diversity. Hidden objects and obstacles must be characterized by studying historical records and conducting geophysical surveys. Some of these include sewers, cables, underground tanks and tunnels, piping, foundations and other infrastructure that must all be identified. Ease of extraction is an important factor. Tanks are much easier to remove than large foundations, which may require blasting. Concrete is more difficult to remove than brick or wood. These are summarized in Figure 6.7.

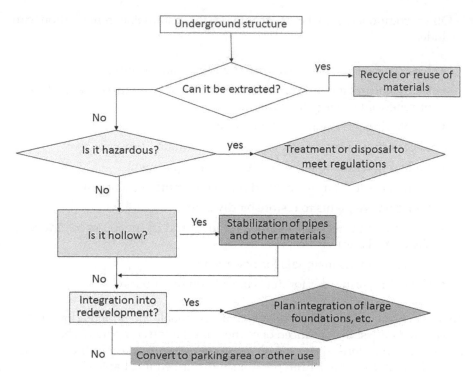

FIGURE 6.7
Managing of underground structures in redevelopment. (Adapted from Genske [2003].)

The USEPA has made efforts since 1995 to encourage brownfield redevelopment—in line with the goals of sustainability (USEPA 2017). More than $22 billion has been leveraged that amounts to about $18 of leverage per EPA brownfield dollar. About 7.3 jobs per $100,000 have been leveraged also, thus totally more than 117,000 jobs in the United States and 70,000 acres have been readied for reuse. The USEPA (1999a) developed a framework shown in Figure 6.8. The framework is designed to assist municipalities, planners, and developers to undertake brownfield projects to enable the project to be sustainable for the community in the future. The principles include resource conservation, materials reuse, public safety and mobility, and information availability. Measures of sustainability should be incorporated and projects should be reevaluated every few years to determine progress and failures.

The USEPA (2017) has indicated that brownfield development in five pilot studies showed that vehicle miles are reduced compared to development of a greenfield by 32%–57%, which means reduced pollutant and greenhouse emissions. Estimates also showed a 47%–62% reduction of stormwater runoff for brownfield site development. Cleaning up brownfield properties led

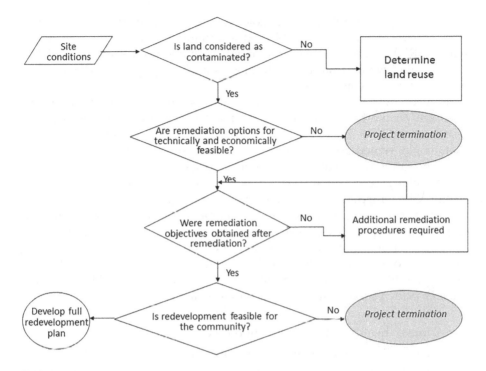

FIGURE 6.8
Framework for brownfield redevelopment. (Adapted from USEPA [1999b].)

to residential property value increases of 5%–15.2% within 1.29 miles of the sites. Local government tax revenues in the year after cleanup increased by seven times the amount that the USEPA contributed to the cleanup of those brownfields ($12.4 million).

In Canada, the Green Municipal Fund of the Federation of Canadian Municipalities supports brownfield remediation of contaminated lands and implantation of renewable energy projects on contaminated land (FCM 2018). Case studies are shown as follows. A residential condominium was built in Ottawa in the Currents) on a former gas station, car wash, and dry cleaner. Redevelopment fees were waived by the city as an incentive. Support was received from the Green Municipal Fund. The project was completed in 2007 after 1.5 years. The main contaminants were hydrocarbons and per-chloroethylene that were removed from the site (about 400 L of water and 4,400 m^3 of soil). Property taxes increased by 1,300% and property assessments by 3,000% due to the redevelopment. Both businesses and residents believed the redevelopment was positive for the community and made the area more vibrant and a better place to live.

The need to incorporate sustainability into projects is increasing due to pressure from all stakeholders. The Network of Industrially Contaminated

Sites in Europe (NICOLE, www.nicole.org/) is also concerned with sustainable remediation and has established a framework for Europe and works with SURF-UK. Sustainable remediation includes implementation of low energy intensive and natural technologies such as phytoremediation, biobarriers, and renewable energies.

SURF US was the first SURF initiative. Sustainable remediation guidelines are on their website (www.sustainableremediation.org/remediation-resources/) (SURF 2009). According to the SURF framework (SURF 2009), sustainable remediation is defined as "sustainable approaches to the investigation, assessment and management (including institutional controls) of potentially contaminated land and groundwater." They indicate that sustainable remediation should minimize energy consumption and use of natural resources, minimize emissions particular to the air, simulate or employ natural processes, reuse and recycle land and materials, and encourage remediation practices that destroy contaminants instead of transferring them to another phase. Other SURF initiatives are in Mexico, Colombia, Brazil, Italy, Netherlands, China, Japan, Taiwan, Australia, and New Zealand. The key principles of sustainable remediation are related to

- Human health and environment protection
- Safe working practices for workers and the local communities
- Decision-making in a clear, consistent, and reproducible manner
- Clear, understandable transparent record keeping and reporting
- Stakeholder involvement in a clear practice
- Decision-making based on scientific relevant and accurate data

The overall framework is summarized in Figure 6.9. Some examples of sustainability considerations include: land use should be matched with land

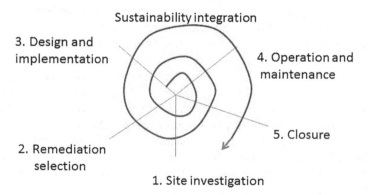

FIGURE 6.9
SURF framework for integrating sustainability throughout the project life cycle (Holland et al. 2011).

TABLE 6.2

Sustainability Indicators for Remediation Evaluation

Environmental Aspect	Economic Aspect	Social Aspect
Air emissions	Capital gain	Community involvement and satisfaction
Ecological and biodiversity impact	Direct economic benefits and costs	Compliance with objectives and strategies
Energy consumption	Financial recovery	Consideration of social sensitivities
Ground and surface water quality	Indirect benefits and costs	Ethical considerations and equity
Intrusiveness	Employment	Human health impact
Impact on habitat	Gearing	Inconvenience
Natural resource use	Life span and risks	Job creation
Soil quality	Litigation potential	Uncertainty and evidence
Sediment quality	Net present value of costs	Worker safety
Waste generation	Project flexibility	
Water use	Property reuse	

Source: Adapted from CL:AIRE (2010).

conditions (contaminant, geotechnical properties) and using vapor barriers instead of removing large volumes of soil, or using sustainable drainage systems for percolation into the clean soil. To evaluate the environmental, social, and economic aspects, the objectives of the assessment need to be determined such as selection of stakeholders, boundaries of the assessment, the type of sustainability indicators, the type of assessment method, and what the sensitivity of the analysis will be. Table 6.2 shows some indicators that can be used for sustainability assessment of remediation options.

6.3 Sustainable Water Treatment

For water treatment, physical–chemical techniques include physical and/or chemical procedures for removal of the contaminants including precipitation, air stripping, ion exchange, reverse osmosis, electrochemical oxidation, etc.

As an example, techniques for groundwater treatment for arsenic are shown in Table 6.3. To evaluate the sustainability of these methods, factors such as materials, energy, transportation, and waste management requirements need to be considered. Although simple ion exchange techniques are used, they are often insufficient for treatment of all species of arsenic. For example, oxidation of As(III) to As(V) is required and performed with

TABLE 6.3

Comparison of Technologies for the Remediation of Arsenic-Contaminated
Groundwater

Technology	Waste Stream	Treatment of Waste	Disposal Options
Activated alumina with regeneration	Alkaline and acidic liquids	Neutralization and precipitation with ferric salts	Sewer, wastes into landfill
Coagulation/ filtration	Ferric sludge, redox sensitive, 97% water content	Dewatering and drying	Landfill after dewatering, brick or other product manufacture
Iron oxide filters	Exhausted adsorbent, redox sensitive, <50% solids, passes TCLP test[a]	No treatment	Landfill, immobilization, brick or other product manufacture
Ion exchange	Liquid saline brine	Precipitation with ferric salts	Sewer, brine discharge, landfill for residual, possible recycling of brines
Lime softening	Backwash water	pH adjustment	Sludge for landfill
Membrane techniques such as reverse osmosis or nanofiltration	Concentrated liquids	None performed	Sewer or brine discharge
Nanotechnology filters	Alkaline solutions	Neutralization	Landfill of precipitates

Source: Adapted from Yong et al. (2014).
[a] TCLP refers to the Toxicity Characteristic Leaching Procedure.

a pre-oxidation filter. This requires disposal of a toxic arsenic waste resulting from the regeneration of these filters impacting the sustainability of the treatment process. Generation of significant quantities of wastes can severely impact the environment and cause more harm than good. Sustainable economic solutions are needed. New combinations of materials as sorbents such as granular activated carbon (GAC) supported nanoscale zero-valent iron (nZVI) are being developed (Chowdhury and Mulligan 2013) and other nanomaterials (such as Fe-Cu) (Babaee et al. 2018). Treatment facilities or alternative water sources are required. A more sustainable approach would reduce arsenic input into the groundwater, if possible.

Traditional wastewater in developed countries is made up of toilet water, gray water from kitchens, showers, clothes washing, industrial wastewater, and stormwater. All are combined and sent to the wastewater treatment facility for treatment, which requires substantial energy. The discharges from these plants contain heavy metals, nutrients, endocrine-disrupting chemicals, and some organic matter. This pollutes the surface water further. Removal of nutrients is expensive and not widely practiced. This type of treatment is quite unsustainable. In addition, large amounts of solids are produced and must be managed.

Effluent water quality is one of the main bases for evaluating treatment capabilities of water treatment processes. Energy use, nutrient addition and recovery, and other requirements need to be included to determine if the process is sustainable. Recycling of resources needs to be practiced as much as possible. The sustainability of the following nitrogen removal systems was evaluated by Mulder (2003): (a) conventional activated sludge systems, (b) an activated sludge system with autotrophic nitrogen removal (combined nitrification and anaerobic ammonia oxidation), (c) duckweed or algal ponds, and (d) engineered wetlands (Figure 6.10). The six sustainability indicators were sludge production, energy and space requirements, recovery of resources, and N_2O emissions. The system of autotrophic nitrogen removal is the most sustainable and no addition of organic matter is required, (b) low sludge production, and (c) high nitrogen removal.

One of the more recent significant developments in water treatments is related to the use of anammox bacteria for nitrogen treatment. Anammox is an abbreviation for anaerobic ammonium oxidation. Nitrite and ammonium ions are converted to nitrogen gas and water via the reaction:

$$NH_4^+ + NO_2^- \rightarrow N_2 + 2H_2O \qquad (6.1)$$

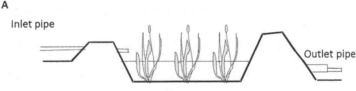

A

Inlet pipe

Outlet pipe

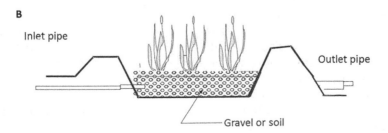

B

Inlet pipe

Outlet pipe

Gravel or soil

FIGURE 6.10
Schematic representation of (A) free surface and (B) subsurface flow constructed wetlands. (Adapted from Mulligan [2002].)

To grow the slow growing bacteria, granular biomass or biofilm systems are most often used in sequencing batch reactors (SBR), moving bed reactors or gas-lift-loop reactors. In 2002, the first full-scale reactor based on the application of anammox bacteria was built in the Netherlands in 2002 (van der Star et al. 2007). Applications have been for municipal wastewater or industrial effluents from a tannery and a potato processing plant.

There is considerable cost reduction potential for the annamox process compared to conventional nitrogen removal. In conventional nitrogen removal, nitrification by aerobic ammonia- and nitrite-oxidizing bacteria is followed by denitrification to convert nitrate to nitrogen. There are substantial requirements for aeration and input of organic substrates (typically methanol). Therefore, these two processes require high amounts of energy, produce excessive sludge, and significant amounts of GHGs such as CO_2 and N_2O and ozone-depleting NO (Hu et al. 2013). However, anammox bacteria can convert ammonium and nitrite directly to N_2 anaerobically, which requires only partial oxidation to nitrite instead of full conversion to nitrate. Less sludge is also produced. The process can reduce CO_2 emissions and costs by up to 60% compared to conventional nitrogen removal. Conversion rate up to 5–10 kg N per m^3 of water (van Loosdrecht 2008) have been obtained. The main disadvantages of the process are the slow doubling time of 10 days to 2 weeks, and the sufficient amounts of sludge for start-up of a reactor, which can be difficult to obtain.

Further developments are related to simultaneous partial nitrification/anammox (PN/Anammox) process in a single-stage reactor configuration to replace two-stage anammox technologies (separate reactors for PN and anammox processes). This reduces the cost and energy requirements for nitrogen removal from wastewater (Chen et al. 2009; Christensson et al. 2013). Other improvements in the PN/anammox process are related to process optimization by controlling low dissolved oxygen (DO) and other design modifications. Balancing DO for aerobic nitrifiers and anoxic anammox communities is a major challenge (Strous et al. 1997).

BioCAST technology is a continuous single-stage multi-zone reactor that has been used for the removal of organic materials, nitrogen, and phosphorous (Alimahmoodi et al. 2012). Nitrogen removal performance and microbial distribution in a pilot-scale partial nitrification/anammox bioreactor (BioCAST) was recently studied. The reactor contained both suspended and attached biomass. The removal of 73% ± 19% at 32°C–35°C and 87% ± 9% at 27°C–29°C of ammonia was obtained from a synthetic wastewater with influent ammonium loading rates of 175–250 g/m^3/day under a low DO concentration of 1.2 mg/L (Saborimanesh et al. 2017).

Fenner and Flores (2011) have proposed a scheme to incorporate sustainability into the wastewater treatment system (Figure 6.11). The scheme is particularly of interest to developing countries and more remote communities in developed countries. The sustainability principles include adaptation to local conditions, conservation of resources, and

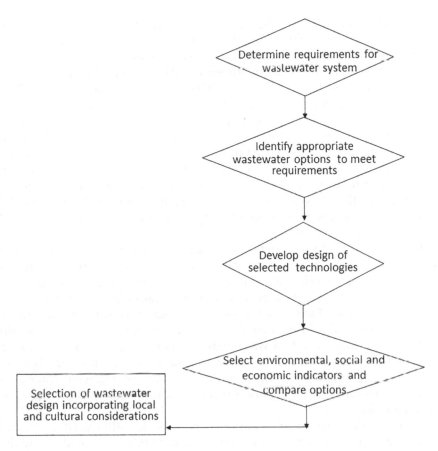

FIGURE 6.11
Development of a sustainable wastewater scheme. (Adapted from Fenner and Flores [2011].)

minimization of waste. Decentralization decreases the pumping required and the need for expensive centralized plants. Resource recovery and stream separation is easier at small scale. Recycling of gray water for toilet flushing is feasible. Industrial wastewaters should be treated separately as they can contain various heavy metals and other chemicals. Stormwater is also highly variable and can contain contaminants from road runoff. Minimization of water use decreases material and energy requirements. Reclamation of water for reuse in industry, agriculture or groundwater recharge water will have to respect environmental norms and thus may need to be treated to achieve these norms. Energy can be recovered from wastewater via anaerobic treatment. This produces methane that has similar properties to natural gas.

Solids from the wastewater treatment plants have often been disposed of in landfills which are becoming highly limited. However, the biosolids containing nutrients, organic matter, and energy could be used in other applications. Incineration is being used more frequently. Energy recovery should be employed if this option is used. Land application is beneficial if contaminant levels such as heavy metals are not elevated.

For example, at a wastewater treatment plant in Canada, the sludge is incinerated and sent for disposal. A study (Mulligan and Sharifi-Nistanak 2016) was performed to determine if the sludge could be used beneficially. High concentrations of cadmium, copper, cobalt, and selenium needed to be reduced to pass regulations before use as a fertilizer. A washing method with a solution of K_2HPO_4 was chosen to preserve the nutrients while removing the heavy metals. The amounts of cadmium, copper, cobalt, and selenium were reduced up to 80%, 44%, 70%, and 93%, respectively. The nutrient concentrations in the biosolids were 17% nitrogen, 17% phosphorus and 25% potassium, respectively. This potentially is a more sustainable approach than incineration of the solids.

The feasibility of CO_2 removal in combination with anaerobic treatment of synthetic wastewater was initially investigated by Alimahmoodi and Mulligan (2008). Subsequently, the feasibility of applying this method for CO_2 removal from industrial emissions with treatment of industrial paper wastewater was evaluated as the pulp and paper industry is significant, especially in Canada. To evaluate this method, industrial emissions containing CO_2 were injected into a wastewater stream where CO_2 is biologically converted to methane as a biogas. Bioconversion of CO_2 to methane is based on the last step of anaerobic digestion (methanogenesis) in which methanogenic archaea convert CO_2, H_2, and simple organic molecules to methane and carbon dioxide.

Consequently, with the addition of carbon dioxide after wastewater pollutant degradation (that provides acetic acid and hydrogen), methane with a high efficiency can be produced through a highly sustainable process. Abedi et al. (2012) determined the effect of pH (6.5, 7.0 and 7.5) and temperature (20°C, 30°C, and 35°C) on CO_2 removal by anaerobic treatment of Kraft and CTMP pulp and paper wastewater. Subsequently, the optimum condition was investigated for CO_2 removal by anaerobic treatment of recycled paper wastewater. COD and carbon dioxide removal in addition to methane generation were determined through batch tests, with and without CO_2 addition. CO_2 can be converted to methane during the wastewater treatment for COD removal (Figure 6.12). Higher temperatures and lower pH values improved removal efficiency.

This method can be applied for carbon dioxide removal with wastewater treatment. It also shows a high efficiency in carbon dioxide reduction. This method is less complex than other methods for CO_2 removal and is promising for mitigating carbon dioxide emission reduction and global warming. The methane generated in this can be used as a source of energy in the plant. Many other wastewater treatment technologies are available and are detailed in Metcalf and Eddy (2013) and other sources of information.

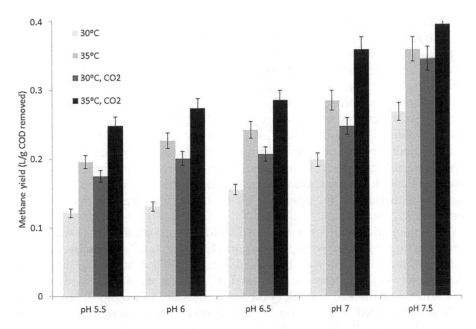

FIGURE 6.12
Treatment of recycled paper wastewater with and without carbon dioxide injection at 30°C and 35°C.

6.3.1 Case Study on the Treatment of Mining Wastewater

The Agnico Eagle Mines Ltd. (AEM)'s LaRonde mine is situated in Cadillac, Quebec, Canada between Val d'Or and Rouyn-Noranda. It is an underground mine 3.1 km below the surface with a crusher, wastewater treatment, crusher and related facilities. Approximately 7,300 metric tons/day of ore are processed. The extraction of gold is performed using cyanide that generates an effluent with cyanide and thiocyanate (SCN^-). Thiocyanate can form from the cyanidation of ores containing sulfides (Gould et al. 2012). The clarified water from the tailings pond is treated by hydrogen peroxide and silicate to remove CN. The increase in pH by lime addition helps to precipitate the metals in solution at the same time. Subsequent oxidation of SCN forms ammonia nitrogen (Gould et al. 2012). Nitrogen removal is necessary due to the potential for eutrophication and aquatic toxicity.

Since the early 2000s, the government imposed toxicity tests for mining effluents. The effluent was toxic and water discharge to the environment was suspended to prevent any damage to the receiving environment. Since water accumulated within the ponds on site, a solution needed to be found. Thiocyanate was the source of the toxicity. The Laronde Mine has been aiming to reduce toxicity and contaminants in their water that contains ammonia, nitrite, cyanide, thiocyanate, cyanate, copper, and zinc. The cyanide process creates a difficult to treat effluent. A toxicity test was used to evaluate an effluent from a gold mine with ammonia, nitrate/nitrite, cyanide, cyanate, dissolved organic carbon, copper, and zinc (Wagner et al. 2002). Certain substances in the water contribute to its toxicity but it was difficult to correlate with contaminant concentration. In addition, the water quality can vary according to the season and mineral treatment process. The objective was for the effluent to be non-acutely lethal (nontoxic) for rainbow trout and *Daphnia magna*. Six treatments were evaluated including GAC+zeolite, air stripping + GAC, zeolite + GAC, multistage GAC, alkaline chlorination+dechlorination, and bentonite-based polymers. Three effluents (one from treated tailings water, treated acid rock drainage, and water from an underground mine) were evaluated. The mixture GAC plus zeolite was the preferred option for toxicity reduction but the operating cost is high. Biological treatment would be lower in cost.

Therefore, nine treatments were evaluated, the previous six plus three others including treatment with peroxide and UV light, GAC and zeolite powder addition, and treatment with bacteria (Grondin 2002). A biological treatment was shown to be effective for removal of cyanide, cyanates, thiocyanates, ammonia, and metals. The pilot process (with rotating biological contactors, RBCs) was started in 2002, and in 2003, it was in the preengineering stage with a start-up in the fall of 2003.

An initial toxicity identification evaluation was used to evaluate the toxicity of the gold mine effluent. An RBC with extensive surface area for attached growth was evaluated for treatment of the effluent. The system was able to reduce treatment costs. Three test columns of 30.5 cm in diameter filled with limestone rock media were then used, two were aerobic and one was anaerobic. The aerobic ones were able to oxidize the cyanide, cyanate, thiocyanate metals and promote ammonia to nitrate. The anaerobic one was successful for conversion of nitrate to nitrogen gas. Toxicity removal from the final effluent was confirmed. A third aerobic column for polishing would be used for polishing the effluent.

A large volume of contaminated water had accumulated in the tailings pond since 2000 since no technology for SCN^- treatment was used. A chemical process was also tested but toxicity inhibited the discharge of the effluent into the environment. After 2 years of research, a biological treatment process for cyanate and thiocyanate was proposed. The water with free cyanide and metals was then fed to the biological water treatment process. Biotechnology was chosen to resolve the problem as toxicity could be eliminated at low cost without dangerous by-product formation. The temperature of operation and the sensitivity of the bacteria to change were challenges.

The first of two lines was constructed in 2003 after pilot tests. A second series of two lines was constructed in 2004. The processes are as follows:

$$SCN^- + 3H_2O + 2O_2 \rightarrow SO_4^{2-} + HCO_3^- + NH_4^+ + H^+ \qquad (6.2)$$

$$NH_4^+ + \frac{3}{2O_2} \rightarrow 2H^+ + H_2O + NO_2^- \tag{6.3}$$

$$NO_2^- + \frac{1}{2}O_2 \rightarrow NO_3^- \tag{6.4}$$

$$NH_4^+ + 2HCO_3^- + 2O_2 \rightarrow NO_3^- + 2CO_2 + 3H_2O \tag{6.5}$$

The RBC process consisted of four propylene disks on steel shafts (Lavoie et al. 2008). The disks had a surface area of 461,000 m² (80% submerged) in aerated basins of 1600 m³. Initially, $14 million were invested and this was followed by $1 million per year. At the beginning, there was an accumulation of nitrite and thus treated effluent was not discharged in the environment and was accumulated in ponds on the mine site. Findings of additional research work showed that four stages were needed to allow degradation of cyanate and thiocyanate to ammonia (*Pseudomonas*), ammonia (*Nitrosomonas*) to nitrite, nitrite to nitrate (*Nitrobacter*). High concentrations of ammonia in the second stage of the process led to inhibition of the nitratation (nitrite to nitrate). An internal recirculation and final water treatment plant enabled the objectives to be obtained. There are still ongoing issues regarding stability due to seasonal variations and breaking shafts. Reduction in operation costs and the achievement of the objectives were studied. The implementation of a denitrification process for the transformation of nitrate to nitrogen gas without addition of a source of carbon was foreseen. The steps to decrease the accumulated water at the mine are shown in Figure 6.13.

Some other issues occurred during the process. Anti-scaling agent in the feed led to inhibited growth of bacteria and precipitation of elements in the feed and heat exchanger tanks were replaced by a non-inhibiting agent. Thiocyanates in the water varied seasonally and as a function of ore composition. Separation of the four lines permitted biomass to develop and recirculation was added to maintain a stable concentration of SCN.

After start-up in 2004, the next two years allowed the adaptation of the bacteria. Covers were modified for the reactors to protect from UV light. Ferro-oxidizing bacteria attacked the iron piping and thus had to be replaced. The anticorrosion agent used to reduce corrosion was inhibitory to the bacteria. Therefore, this product was replaced. The higher the temperature, the better the performance. Thus, the increase of the hot water heater power increased the performance of the bacteria. The concentration of the oxygen was too low and thus the injectors were not large enough. Too high concentrations of thiocyanate or ammonia inhibited the bacteria downstream; thus, dilution was necessary. Also, phosphorus had to be added. Although it was hoped to add only the quantity needed, an excess was required. As the phosphorus had to be eliminated, an alum addition was made to remove the phosphorus.

An RBC was used for many years but many mechanical failures occurred due to the weight of accumulated biomass that did not slough off. Therefore,

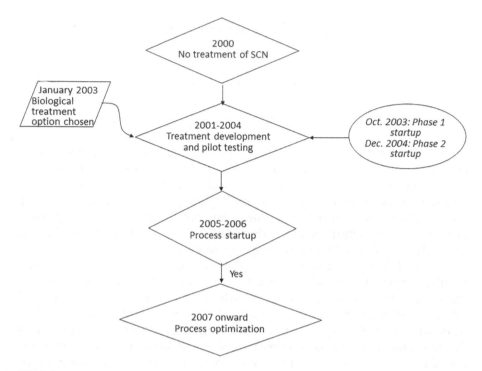

FIGURE 6.13
Biological process for mining effluent startup.

AEM partnered with Headworks BIO Inc. (2012) (www.headworksbio.com) to install the moving bed biofilm bioreactor (MBBR) (Figure 6.14). Pilot tests were carried out over 4 months to treat 450 mg/L of thiocyanate. The system acclimated for 2 weeks. This was followed by 20 days of commissioning before operation was initiated. Flows were increased from 35 to 45 m³/h and then to the normal flow of 65 m³/h. The hydraulic retention time (HRT) was 3.7 h. The retrofit involved a fluidized bed that enables self-sloughing, self-regulated biofilm, and no mechanical components; therefore, no mechanical failure would occur. The Laronde gold mine had four trains of four-stage RBCs for the treatment of SCN⁻ and OCN⁻ in wastewater. Two of the first-stage RBC reactors were replaced by MBBRs. In the first stage, these components are converted to ammonia, sulfate, and carbon dioxide and then nitrification occurs in the next three reactors. The high ammonia concentration can be detrimental to these later stages. Internal dilution was performed by recycling the effluent from the fourth stage to the second-stage reactor.

Therefore, more R&D was needed and a budget was thus allocated. The MBBR has a 687,000 m² surface area compared to 310,000 m² surface area for an RBC. The research consisted of optimization of pH control, and pilot testing of media for nitrification and denitrification. The MBBR was implemented

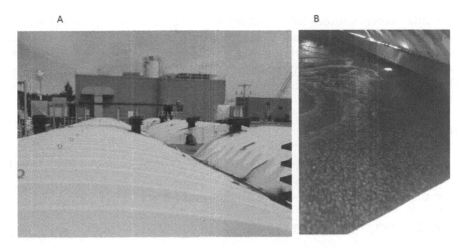

FIGURE 6.14
(A) Covered tanks for biological treatment and (B) Interior of the moving bed biofilm bioreactor.

in 2007 and flow increased to the objective of 250 m³/h. The MBBR retrofit provided an increase in capacity without interrupting mining operations. Due to the overload of phosphorus, in 2007 a phosphorus removal plant was implemented ($1 million project). A level of 85% removal was achieved by coagulation/precipitation.

The water in the basin reached 4.5 million m³ by 2009. The treatment plant thus had to be modified. The MBBR system was implemented and the ammonia treatment was modified to allow all year treatment. There were many problems with the latter system. Therefore, two reactors were transformed to nitrification. Ammonia was reduced and nitrate transformation increased. The volume of water in the pond was reduced. In 2013, another reactor broke and was transformed to an MBBR. In 2014, 2.1 million m³ of water were treated. The start-up of the bacteria was long (5–8 months) before the biofilm was produced. They were very sensitive to metals and new products. A protocol was established to maintain the performance. The MBBR performed twice as well as the RBC. The nitrifiers performed well and thus two other reactors were modified. The ammonia treatment plant was stopped in 2014. The long-term objective was to maintain the water in the pond at 1.5 million m³.

Prenitrification is a strategy to promote removal of nitrogen compounds in the presence of organic contaminants SCN⁻ and OCN⁻ (Villemur et al. 2015). Lab-scale reactors were established to evaluate the prenitrification step feasibility (Villemur et al. 2015) using a synthetic effluent. Two four-stage moving MBBRs (designated as A and B trains) were implemented for the treatment of thiocyanate, cyanate, and ammonia. Both trains removed all three components entirely. The first stage of the B train consisted of a prenitrification step with cyanate and thiocyanate as the only carbon sources thus resulted in a higher total N removal (62.6% compared to 38.5%). *Thiobacillus* spp. was dominant in all stages, and *Nitrobacter, Nitrosospira, Nitrosomonas,* and *Nitrospira* were found in the second and third stages. Annamox bacteria likely also were active in the train B nitrogen dissimilatory process. This aspect could be further exploited in the future.

To complete the removal of nitrogen a denitrification step was added to the process.

$$NO_3^- \rightarrow NO_2^- \rightarrow NO \rightarrow N_2O \rightarrow N_2 \qquad (6.6)$$

Overall the reaction with the thiocyanates and nitrates is as follows:

$$5SCN^- + NO_3^- + HCO_3^- + H_2O + 8H^+ \rightarrow 5SO_4^{2-} + 5NH_3 + 10CO_2 + 4N_2 \quad (6.7)$$

The first reactors could be anaerobic. This would allow the nitrates to be brought to the front, and the bacteria that transform the thiocyanate could use the nitrate for oxygen in anaerobic conditions. A test was done to confirm the potential. A pilot system was to be tested. Another option is to put the denitrification step early in the process and yet another is to add methanol at the end with the nitrites instead of nitrates. The last option is denitrification in the new anaerobic reactors with methanol addition. Twelve new reactors would be needed. Pilot tests were to be performed.

The cost of the biological treatment plant is \$1.46 per m^3 water treated (24% operation, 39% chemical, and 34% energy from electricity and gas). Electricity was used mainly for the pumps and aerators. Gas was used for the water heater. Hot water from the mine could be used to heat the water in the process. Chemicals in the form of sodium carbonate were added. Costs of operation need to be reduced, particularly by reuse of available energy.

Nitrification–denitrification was subsequently evaluated for a gold mine effluent with a concentration of SCN$^-$ of 435 ± 53 mg/L and NH$_3$-N of 41 ± 9 mg/L. The HRTs in the biological reactors are 6h each (Tanabene et al. 2018). In the first, the SCN$^-$ and OCN$^-$ are oxidized to sulfate, carbon dioxide and ammonia, then oxidized to nitrite and nitrate. *Nitrobacter, Nitrospira, Nitrosomonas, Nitrospira,* and *Thiobacilli* were found in the latter reactors. Pilot tests were conducted at 20 L scale as shown in Figure 6.15. The sequence is pre-denitrification, then oxidation, and then methanol addition for post-denitrification. SCN$^-$ and NO$_2^-$ removed at >97 and >70%, respectively, using an HRT of 5.5h and a dilution factor of 1.5. Nitrate was removed at >80% using 0.18 g methanol per gram of nitrate. The process must be improved for the northern climate and reduction of water heating is needed. In summary, this case study provides an example of ongoing innovation to resolve difficult to treat effluents in a cost-effective and energy-efficient manner.

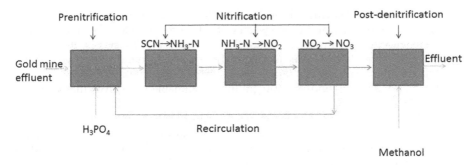

FIGURE 6.15
Schema of pilot nitrification–denitrification system. (Adapted from Tanabene et al. [2018].)

6.4 Conclusions

In the 1990s, intensive treatment processes were the accepted methods once contaminant levels exceed regulatory levels. However, incorporation of reuse and waste reduction is now increasingly employed to address sustainability goals. Reduction of contaminant levels should not be the only consideration for soil or water remediation. Innovative approaches are needed to perform the treatment in a sustainable manner as possible. Energy requirements should be reduced and renewable energies should be employed more frequently to reduce GHG emissions. Waste generation should be reduced and resources should be protected. Consultation with stakeholders should be performed to incorporate environmental, social, and economic concerns into the remediation. Tools are being developed and employed to enable comparison of the sustainability of technologies and provide a transparent means of communication of the results. There are guidance documents such as from SURF initiatives for site remediation, but there are no universally accepted methods for sustainable evaluations. Sustainability incorporation should be across the entire project life cycle including the evaluation of the contamination, development of alternatives, process design, implementation and operation, and decommissioning of the process.

References

Abedi N., X. Jiang, C.N. Mulligan and L. Yerushalmi, 2012. Development of a sustainable method to reduce carbon dioxide emissions by bioconversion into methane. In: *CSCE Conference*, Edmonton, AB. Accessed June 6–9, 2012.

Alimahmoodi, M. and C.N. Mulligan, 2008. Anaerobic bioconversion of carbon dioxide to biogas in an upflow anaerobic sludge blanket reactor. *Journal of Air and Waste Management Association*, 58: 95–103.

Alimahmoodi, M., L. Yerushalmi and C.N. Mulligan, 2012. Development of biofilm on geotextile in a new multi-zone wastewater treatment system for simultaneous removal of COD, nitrogen and phosphorus. *Bioresource Technology*, 107: 78–86.

ASTM, 1998. Standard Guide for Remediation of Groundwater by Natural Attenuation at Petroleum Release Sites, ASTM Designation, E1943-98.

ASTM, 2013. Standard Guide for Integrating Sustainable Objectives into Cleanup, E2876-13, June 2013.

ASTM, 2016. Standard Guide for Greener Cleanups, E2893-16, May 2016.

Babaee, Y., C.N. Mulligan and S. Rahaman, 2018. Removal of arsenic (III) and arsenic (V) from aqueous solutions through adsorption by Fe/Cu nanoparticles. *Journal of Chemical Technology and Biotechnology*, 93(1): 63–71.

Chen, H., S. Liu, F. Yang, Y. Xue and T. Wang, 2009. The development of simultaneous partial nitrification, ANAMMOX and denitrification (SNAD) process in a single reactor for nitrogen removal. *Bioresource Technology*, 100: 1548–1554.

Chowdhury, R. and C.N. Mulligan, 2013. Removal of arsenate from contaminated water by granular carbon embedded with nano scale zero-valent iron. In: *CGS Annual Conference*, Montreal, QC, September 29–October 2, 2013.

Christensson, M., S. Ekström, A.A. Chan, E. Vaillant Le and R. Lemaire, 2013. Experience from start-ups of the first ANITA Mox plants. *Water Science & Technology*, 67: 2677–2684.

CL:AIRE, 2010. *A Framework for Assessing the Sustainability of Soil and Groundwater Remediation, SuRF Sustainable Remediation Forum*, London, CL:AIRE.

FCM, 2018. Green Municipal Fund. Case Studies: Taking Action of Brownfields. https://fcm.ca/Documents/brochures/LiBRe/Taking_Action_On_Brownfields_EN.pdf. Accessed June 6, 2018.

Fenner, R. and A. Flores, 2011. Providing sustainable sanitation. In: *Sustainable Development in Practice: Case Studies for Engineers and Scientists*, 2nd edition, A. Azapagic and S. Perdan (eds), West Sussex, UK: John Wiley & Sons, pp. 348–373.

Genske, D.D., 2003. *Urban Land, Degradation, Investigation and Remediation*, Berlin: Springer, 331 pages.

Golder Associates, 2018. Goldset. https://golder.goldset.com/portal/toolcustomization.aspx. Accessed November 13, 2018.

Gould, W.D., M. King, B.R. Mohapatra, R.A. Cameron, A. Kapoor and D.W. Koren, 2012. A critical review on destruction of thiocyanate in mining effluents. *Minerals Engineering*, 34: 38–47.

Grondin, L., 2002. Identification de la toxicité à l'effluent de la Mine Laronde. 1st edition of *Mines and Environment Symposium*, Rouyn-Noranda, QC, November 3–5, 2002, 7 pages.

Headworks BIO, 2012. Case Study: Agnico Eagle LaRonde Gold Mine. Headworks.com.

Holland, K.S., R.E. Lewis, K. Tipton, S. Karnis, C. Dona, E. Petrovskis, L.P. Bull, D. Taege and C. Hook, 2011. SURF framework for integrating sustainability in to remediation projects. *Remediation*, 21: 7–38.

Hu Z., T. Lotti, T. Lotti, M. de Kreuk, R. Kleerebezem, M. van Loosdrecht, J. Kruit, M.S.M. Jetten and B. Kartal, 2013. Nitrogen removal by a nitritation-anammox bioreactor at low temperature. *Applied and Environmental Microbiology*, 79(8): 2807–2812.

ISO, 2006. ISO/DIS 14044, *Environmental Management – Life Cycle Assessment – Requirements and Guidelines*, Geneva, Switzerland: ISO.

Laporte, P., 2015. Évolution à la mine Laronde du traitement biologique du thiocyanate. 5th edition of *Mines and Environment Symposium*, Rouyn-Noranda, QC, June 14–17, 2015, 6 pages.

Lavoie, P., S. Bérubé, V. Bougie and P. Juteau, 2008. La mise en route d'une usine biologique dans le domaine minier. 3rd edition of the *Mines and Environment Symposium*, Rouyn-Noranda, QC, November 3–5, 2008, 31 pages.

Metcalf and Eddy Inc, 2013. *Wastewater Engineering: Treatment and Resource Recovery*, 5th edition, Dubuque, IA: McGraw-Hill, 2048 pages.

Mulder, A., 2003. The quest for sustainable nitrogen removal technologies. *Water Science & Technology*, 48(1): 67–75.

Mulligan, C.N., 2014. Enhancement of remediation technologies with biosurfactants. In: *Biosurfactants: Research Trends and Applications*, C.N. Mulligan, S.K. Sharma and A. Mudhoo (eds), Boca Raton, FL: CRC Press, pp. 231–276.

Mulligan C.N., 2002. *Environmental Biotreatment*, Rockville, MD: Government Institutes.

Mulligan C.N. and M. Sharifi-Nistanak, 2016. Treatment of sludge from a wastewater treatment. *International Journal of GEOMATE*, 11(23): 2194–2199.

National Round Table for the Environment and Economy (NRTEE), 2004. *Cleaning up the Past, Building the Future, A National Brownfield Redevelopment Strategy for Canada*, Ottawa, ON: NRTEE.

Saborimanesh, N., E. Castillo Arriagada, D. Walsh, L. Yerushalmi and C.N. Mulligan, 2017. Biological treatment of ammonia-rich wastewater by partial nitrification/ANAMMOX in the BioCAST reactor. In: *14th IWA Leading Edge Conference on Water and Wastewater Technologies*, Florianopolis, Brazil, May 29–June 2, 2017.

Strous, M., E. van Gerven, P. Zheng, J.G. Kuenen and M.S.M. Jetten, 1997. Ammonium removal from concentrated waste streams with the anaerobic ammonium oxidation (anammox) process in different reactor configurations. *Water Research*, 31: 1955–1962.

Sustainable Remediation Forum (SURF), 2009. Integrating sustainable principles, practices, and metrics into remediation projects. *Remediation Journal*, 19(3): 5–114.

Tanabene, R., T. Genty, C. Gonzalez-Merchan, B. Bussière, R. Potvin and C.M. Neculita, 2018. Nitrification-denitrification of thiocyanate, ammonia and nitrates in highly contaminated gold mine effluents using methanol as energy source. *Journal of Environmental Engineering*, 144(5): 05018002 (1–10).

USEPA, 1999a. A Sustainable Brownfields Model Framework. United States Environmental Protection Agency, Office of Solid Waste and Emergency Response, Washington, DC: EPA, EPA500-R-99-001. January 1999.

USEPA, 1999b. Road Map to Understanding Innovative Technology Options for Brownfields Investigation and Cleanup, 2nd edition, United States Environmental Protection Agency, Office of Solid Waste and Emergency Response, Washington, DC: EPA, 542 B-99-009.

USEPA, 1999c. Use of Monitored Natural Attenuation at Superfund, RCRA Corrective Action, and Underground Storage Tank Sites, Information on OSWER Directive 9200.4–17P, USEPA-540-R-99-009.

USEPA, 2009. Green Remediation: Integrating Sustainability Environmental Practices into Remediation of Contaminated Sites. https://19january2017snapshot.epa.gov/remedytech/green-remediation-incorporating-sustainable-environmental-practices-remediation_.html. Accessed November 12, 2018.

USEPA, 2017. Brownfields Program Accomplishments and Benefits. www.epa.gov/
brownfields/brownfields-program-accomplishments-and-benefits.   Accessed
June 1, 2018.

van der Star, W.R.L., W.R. Abma, D. Blommers, J.-W. Mulder, T. Tokutomi, M. Strous,
C. Picioreanu and M.C.M. van Loosdrecht, 2007. Startup of reactors for anoxic
ammonium oxidation: Experiences from the first full-scale anammox reactor in
Rotterdam. *Water Research*, 41(18): 4149–4163.

van Loosdrecht, M.C.M., 2008. Innovative nitrogen removal. In: *Biological Wastewater
Treatment: Principles, Modelling and Design*, M. Henze, M.C.M. van Loosdrecht,
G.A. Ekama and D. Brdjanovic (eds), London: IWA Publishing, pp. 139–155.

Villemur, R., P. Juteau, V. Bougie, J. Ménard and E. Déziel, 2015. Development of four-
stage moving bed biofil reactor train with a pre-nitrification configuration for
the removal of thiocyanate and cyanate. *Bioresource Technology*, 181: 254–262

Wagner, R., L. Liu and L. Grondin, 2002. Toxicity treatment evaluation of mine final
effluent-using chemical and physical treatment methods, SGS Mineral Services,
Technical Bulletin, December 12, 2002, 6 pages. www.sgs.com/~/media/
Global/Documents/Technical%20Documents/SGS%20Technical%20Papers/
SGS%20MIN%20TP2002%2012%20Toxicity%20Treatment%20of%20Mine%20
Effluent.pdf. Accessed July 24, 2018.

Yong, R.N. and C.N. Mulligan, 2004. *Natural Attenuation of Contaminants in Soils*, Boca
Raton, FL: Lewis Publishers, 319 pages.

Yong, R.N. and C.N. Mulligan and M. Fukue, 2014. *Sustainable Practices in
Geoenvironmental Engineering*, 2nd edition, Boca Raton, FL: CRC Press.

7

Indicators for Sustainable Design

7.1 Introduction

As previously seen, sustainability incorporates environmental, economic, and social aspects. In 1993, the UN Conference on Environment and Development, commonly called the Rio Conference was held. Agenda 21 was conceived to achieve sustainable development. The UN Commission on Sustainable Development (UNCSD) produced a set of sustainable development indicators (SDIs) with international agreement.

In 1997, CERES with United Nations Environmental Programme (UNEP) put forward the Global Reporting Initiative (GRI). More than 5,500 corporations use these. GRI has provided substantial information on the selection, and the use of sustainability indicators (www.globalreporting.org). Indicators are of particular use as the concept of sustainability is vague and progress is difficult to measure. In 2002, nine international banks and the International Finance Corporation developed a framework for environmental and social risk for project financing called the Equator Principles. Almost all projects for the developing world are influenced by these rules. A concept that came out of this was the triple bottom line where decision-making must include social and environmental costs.

Numerous organizations were involved in this development, such as SDIs including the European Environment Agency (EEA), United Nations Development Programme (UNDP), The World Bank, World Watch Institute, International Institute of Sustainable Development (IISD), New Economics Foundation (NEF), United Nations Commission for Sustainable Development (UNCSD), WTO (World Tourism Organization) and nationally the Department of Culture Media and Sport (DCMS), and the Department for Environment Transport and the Regions (DETR).

SDIs can be categorized into many ways. These can be viewed as a system at various scales (Sikdar 2003): global, national/regional, business/institutional, and technological (Figure 7.1). Engineers and scientists have the largest influence at the technological scale. They can inform others at other scales but do not have the ability to make decisions. Various indicators and tools have been developed and will be discussed in this chapter.

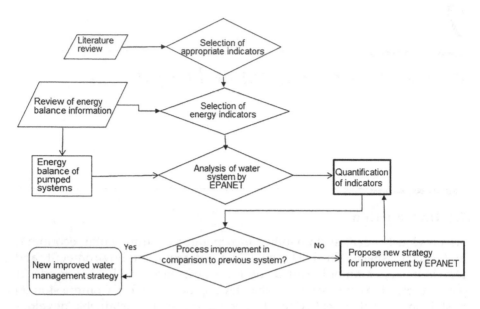

FIGURE 7.1
Classification of sustainability systems.

7.2 Indicators for Assessment

To enable evaluation of progress, indicators are needed. The three domains of sustainability are evaluated to determine improvement. Sikdar et al. (2017) indicated that: "A process or product systems can be made more sustainable by improving the overall impacts of the system (economic, environmental and societal) using quantifiable indicators (or metrics) compared to a reference system with similar attributes." Innovation is needed to achieve the improvement. Progress in one or more of the domains should also be followed over time. To do this, indicators are used. Sikdar et al. (2017) defined indicators as "a quantified measure of the system."

An indicator should be chosen that would enable decision-making in an objective manner. The attribute should be unique and not covered in other indicators. Indicators should be the same for systems that will be compared. They should be transparent to allow public engagement.

In the three pillars of sustainability, environmental, economic, and social, indicators can be selected in one of the domains (1D) at the intersection of two domains (2D) or three domains (3D) as shown in Figure 7.2 (Sikdar 2003). The 1D indicator can also be normalized such as water consumption per capita or per dollar of GDP. Indicators that are only environmental or social are one-dimensional. Those such as socioeconomic indicators cross two dimensions. Others such as energy intensity are the overlap of all three dimensions.

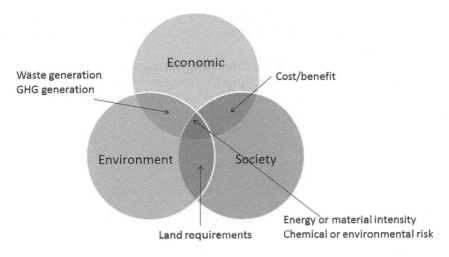

FIGURE 7.2
One-, two-, and three-dimensional indicators. (Adapted from Sikdar et al. [2017].)

At the global level, the Millennium Development Goals (MDGs) set the bar for moving to sustainability. Nations, provinces/states and even cities/towns, etc. have used indicators to show progress to sustainability. Sustainability requires integrated multidimensional indicators linking economy, environment, and society. Good SDIs should be clear, effective, efficient, and inclusive of the needs of stakeholders.

Agenda 21 of UNCSD has indicated that indicators are needed for decision-making. For measuring the sustainability of nations, Sikdar et al. (2017) have indicated that some attempts have been made such as MDG indicators, human development (HD) indicators, material and energy flow analysis-based indicators, or ecological footprint.

For example, for the MDG, indicators for environmental sustainability include:

- Proportion of land covered by forest
- Total carbon dioxide emissions per capita
- Total carbon dioxide emissions per $GDP
- Ozone-depleting substance consumption
- Fish stock proportion within safe biological limits
- Proportion of total resources used
- Proportion of terrestrial and marine area protected
- Proportion of species threatened with extinction
- Proportion of population using an improved drinking water source
- Proportion of population using improved sanitation facilities
- Proportion of urban population living in slums

These indicators are best used at the global scale. Some of the information can be difficult to obtain. The UNEP now uses sustainable development goal (SDG) indicators instead.

At the business or industrial scale, which is the most relevant for engineers, sustainability assessments can be used to identify potential improvements in a product or process on a one time or continual basis. Indicators can be based on (Sikdar et al. 2017):

- Energy
- Water
- Materials
- Toxic releases
- Waste
- Cost
- Worker safety
- Adverse social effects
- Ecological impacts

The usual practice for selection is based on knowledge of the system to be measured. A life cycle point of view should be used to evaluate the production, use, and disposal stages in particular. Indicators can be used to show reduction in water use, emissions, waste, or energy. For example, zero waste generation or 100% renewable energy use can be corporate goals.

Fiksel (2009) indicated the following criteria for selection of sustainability performance indicators:

- Relevance
- Meaningfulness
- Objectiveness
- Effectiveness
- Comprehensiveness
- Consistency
- Practicality

The National Research Council (NRC 2011) Green Book added that the indicators should be actionable, transferrable, intergenerational, and durable. The most effective approach is to use a small number of quantifiable key performance indicators (KPIs). Some indicators according to the UN Sustainability Goals for energy and waste are summarized in Table 7.1.

The Association for Canadian Engineering Companies (ACEC 2016) summarized six key issues for engineers for sustainability. These include water,

TABLE 7.1

Targets of the 2030 Agenda for Sustainable Development and UN Sustainability Indicators for Energy and Waste (UN 2018)

Goal	Indicators
GOAL 7: Ensure access to affordable, reliable, sustainable, and modern energy for all	• Proportion of population with access to electricity, by urban/rural (%) • Proportion of population with primary reliance on clean fuels and technology (%) • Renewable energy share in the total final energy consumption (%) • Energy intensity level of primary energy (megajoules per constant 2011 purchasing power parity GDP)
GOAL 11: Make cities and human settlements inclusive, safe, resilient, and sustainable	• Municipal Solid Waste collection coverage by cities (%) • Annual mean levels of fine particulate matter in cities, urban population (micrograms per cubic meter)
GOAL 12: Ensure sustainable consumption and production patterns	• Compliance with the Basel Convention on hazardous waste and other chemicals • Compliance with the Montreal Protocol on hazardous waste and other chemicals • Compliance with the Rotterdam Convention on hazardous waste and other chemicals • Compliance with the Stockholm Convention on hazardous waste and other chemicals • National recycling rate, tons of material recycled • Number of companies producing sustainability reports

energy, materials, environment, health and safety, and human rights. Human rights covers various aspects such as consideration of cultural heritage, protection of food availability, minimization of corruption and nuisances, such as dust, light, noise, and odor.

Engineers Canada (2016) has released a national guideline on sustainable development for engineers. They have indicated that indicators should be used as early in the process as possible and involve stakeholders in the discussion on the indicators. Appropriate tools, standards, and data should be used as objective evidence. ISO 14000 standards, in particular, should be considered for performance indicators. The indicators should be as measurable, objective, and comparable as possible to enable their use over the life cycle of the project. If similar projects are available, the same indicators should be used and the experience gained from the previous project should be obtained. The type and scope of indicators should be appropriate for the project. Financial needs should be included for the reporting process. Best practices should be followed that includes up-to-date tools and sustainability indicators. Social and economic indicators are very important for sustainability. Quality of life, concerns with noise, light, odor, traditional and cultural values must be considered. An example of the indicators for a city is shown in Figure 7.3.

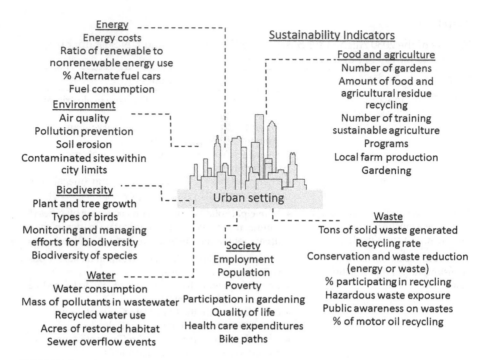

Energy
Energy costs
Ratio of renewable to
nonrenewable energy use
% Alternate fuel cars
Fuel consumption

Environment
Air quality
Pollution prevention
Soil erosion
Contaminated sites within
city limits

Biodiversity
Plant and tree growth
Types of birds
Monitoring and managing
efforts for biodiversity
Biodiversity of species

Water
Water consumption
Mass of pollutants in wastewater
Recycled water use
Acres of restored habitat
Sewer overflow events

Sustainability Indicators

Food and agriculture
Number of gardens
Amount of food and
agricultural residue
recycling
Number of training
sustainable agriculture
Programs
Local farm production
Gardening

Urban setting

Society
Employment
Population
Poverty
Participation in gardening
Quality of life
Health care expenditures
Bike paths

Waste
Tons of solid waste generated
Recycling rate
Conservation and waste reduction
(energy or waste)
% participating in recycling
Hazardous waste exposure
Public awareness on wastes
% of motor oil recycling

FIGURE 7.3
Selected sustainability indicators for a community.

Frameworks for indicators can be goal-based, causal and domain, sector or issue-based (Pinter 2004). Goal-based indicators are to achieve a sustainability goal. The causal framework examines the relationship between indicators to enable prediction of policy responses. Domain-based indicators of concern for sustainability are along the lines of environment, society, and economy such as waste generation and water use.

The 3M has been tracking various indicators since 2005, some since 2002, but the majority since 2012 (3M Sustainability Report 2018) as shown in Table 7.2. Their indicators follow their mission of improving lives. They have set goals for 2025 for reducing greenhouse gas (GHG) emissions by 50% of CO_2e compared to 2002, improving energy efficiency indexed to net sales by 30%, increasing renewable energy to 35% of the total electricity use, reducing global manufacturing waste indexed to sales by 10% and global water use by 10%, and achieving zero landfill status at 30% of its manufacturing sites. Other goals are related to training five million people and reengaging 100% of communities in water stressed—water scarce areas. It can be seen that new metrics are added over time and are sometimes the basis for reporting changes. While some targets are close to being achieved, others such as reducing water use are far.

Companies such as Shell use the GRI reporting standards. They have indicated in their sustainability report that in 2016 (Shell Global 2017) they reduced direct GHG emissions from their facilities to 70 million metric tons on

TABLE 7.2

Indicators for Sustainability by 3M and Progress over the Years

Indicator	Metric Unit	2005	2016	Change %	% change Indexed to Sales
Environmental					
GHG emissions (direct)	Metric tons CO_2 equivalent	10,100,000	4,140,000	59.0	10
Total VOC emissions	Metric tons	6,800	4,630	−31.9	−23
Total energy use	MWh	8,170,000	8,490,000	3.92	2.69
Waste disposal (landfill, treatment, incineration)	Metric tons	145,000	160,000	10.3	16.4
Total hazardous waste	Metric tons	47,700	43,900	−7.97	
Total nonhazardous waste	Metric tons	49.0	45.9	−6.33	
		2012	**2016**		
Waste to energy	Metric tons	42,400	61,000	43.9	
Onsite recycle and reuse	Metric tons	51,100	56,500	10.6	
Total water use	Million cubic meters	43.4	45.9	5.76	4.83
Social		**2012**	**2016**	**Change (%)**	
Full-time employees	Number	84,100	90,100	7.13	
Part-time employees	Number	2,430	2,400	−4.76	
Percentage of female in management positions	Number	24.6	27.8	13.0	
Financial		**2012**	**2016**		
Net sales	Million US$	29,900	30,100	0.669	
Operating income	Million US$	6,480	7,220	11.4	
R&D and related expenses	Million US$	1,630	1,740	6.75	
Capital expenditures	Million US$	1,480	1,420	−4.05	
Cash donations	Million US$	27.9	34.8	24.7	

Source: Data from 3M Sustainability Report (2017).

a CO_2-equivalent basis, flaring from 11.8 million metric tons CO_2-equivalent in 2015 to 7.6 million metric tons in 2016 and the number of spills from 108 in 2015 to 71 in 2016. However, fresh water requirement in 2016 increased to 195 million cubic meters from 186 million cubic meters in 2015, mainly due to oil sands operations. Around 65% of the fresh water consumption was for manufacturing oil products and chemicals and a further 22% was used by oil sands mining operations. Sulfur oxide emissions decreased in 2016 compared to 2015, but nitrogen oxide emissions increased to 122,000 metric tons in 2016 from 104,000 metric tons in 2015 due to inclusion of additional facilities.

VOC emissions increased to 146,000 metric tons in 2016 from 125,000 metric tons in 2015 due to an increase of venting at some facilities. In the future, they expect VOC emissions to decrease with reduced flaring. Regarding waste generation, nine of their downstream manufacturing sites sent more than 50% of it for recycling or reuse. Five of the nine sites sent over 80% for recycling and reuse. An external review committee was put in place to review the report.

To perform a sustainability assessment, the following framework was suggested by Sikdar et al. (2017):

- Identification of the system
- Definition of the scale and system boundaries
- Identification of the appropriate indicators to the system
- Identification of the indicator dimensionality using a Venn diagram
- Collection of the data and calculation of values using appropriate tools
- Determination of the relative indicator weights
- Determination of the system performance
- Determination of sustainability improvements

Many companies use claims of greenness or sustainability by using one or few indicators. For example, electrical vehicles are deemed more environmentally friendly. However, the electricity is often generated by coal power plants. Therefore, there is only a shift in the pollutant production. Similarly, biofuel production requires substantial pollutant production due to agricultural practices and use of clean water for irrigation and land use for biomass production. In addition, many indicators are not verified by a third party. Ecolabeling by best practices such as ISO 14024, Energy Star (www.energystar.gov), or Water Sense (www.epa.gov/watersense) are third party verified.

A sustainability assessment involves determination of environmental, social, and economic impacts and how to reduce them. Analytical tools can be useful for the sustainability assessment such as:

- Environmental impact assessment
- Process design, simulation, and integration of environmental impacts
- Material design to examine if materials are less toxic or waste generating
- Life cycle assessment (LCA)
- Process tools to avoid toxic products and produce more efficiently such as catalysis, biotechnology, or nanotechnology
- Economic tools such as life cycle cost analysis, total cost accounting, and ecological evaluation

The Environmental Protection Agency (EPA) has devised a framework for sustainability indicators (USEPA 2012). Although it was devised for internal use at the EPA, it is useful for other organizations and includes four aspects: public reporting, decision-making, research planning, and program evaluation. In doing the evaluation, the goals needed to be defined, what indicators will be used to evaluate the system, and what unit of measurement (metric) of the indicators will be used, followed by data collection with the appropriated quality assurance procedures and communication of the results to stakeholders as shown in Figure 7.4. The process can be iterative to support decision-making.

There have been many initiatives (more than 900 according the Institute of Sustainable Development). Some are very well known, such as the World Bank and the Organization for Economic Cooperation and Development (OECD). Many indicators are duplicated and many have no supporting documentation. The EPA database (DOSII) is stored as an Excel 2007 workbook and is searchable. It is intended as a repository of indicators. It can be obtained from the website www.epa.gov/research/environmental-tools-support-sustainable-decision-making.

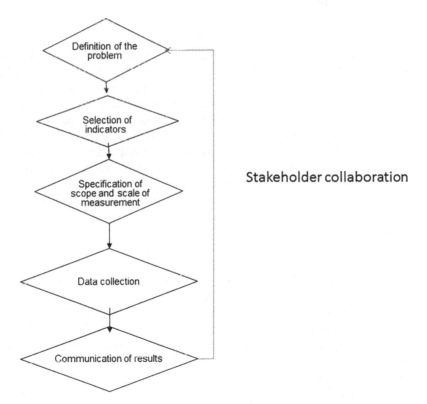

FIGURE 7.4
Implementation of a sustainability assessment. (Adapted from USEPA [2012].)

7.3 Environmental Aspects and Indicators

ASTM International (2015) has developed a guide for evaluating the environmental aspects of sustainability for manufacturing process (E2986-15). Guidance for indicators was provided regarding the selection of indicators. The objective of the evaluation needs to be kept in mind. Characteristics of the indicator are its name, definition, measurement type, unit of measure, reference, and application level.

The American Institute of Chemical Engineers (Beloff et al. 2001) has created a list of six indicators for sustainability metrics that are only in the environmental domain:

- Energy use
- GHG emissions
- Material use
- Pollutant emission
- Toxic emissions
- Water use

The EPA has developed a number of indicators for environmental protection (USEPA 2017). More recently, due to increased attention on sustainability, economic and social factors are considered. The environmental indicators are the report on the environment (ROE) indicators. Consumption indicators have been added that address economic and population factors. These include energy use, freshwater withdrawals, municipal solid waste, and hazardous waste. They, however, are not useful for indicating if practices are moving to sustainability. They are more relevant for coupling to economic and population growth (USEPA 2017). They can help in some decision-making though for balancing human health, environmental, social, and economic risks. Indicators are shown in Figure 7.5 for nutrient discharge.

Trends can be searched for at the website (https://cfpub.epa.gov/roe/indicators.cfm).

Allen and Shonnard (2012) examined environmental indicators for engineering design along the lines of LCA. For example, material and energy inputs and outputs can be considered at each stage as in Figure 7.6. Midpoint or endpoints can be considered as indicators. However, midpoint indicators are less influenced by interpretation. Social and economic impacts should also be considered.

In LCA, as previously discussed, inventories are determined in the life cycle inventory (LCI) stage. Raw materials and emissions are determined. Characterization of the effect on the environment would then need to be based on scientific data or models. For example, methane emissions are

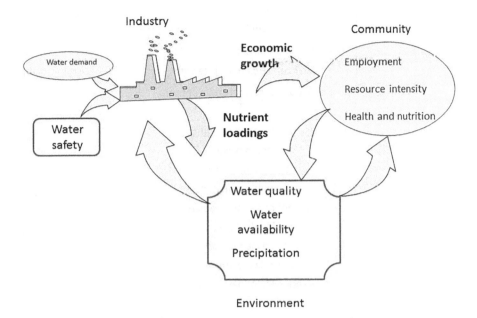

FIGURE 7.5
Indicators of nutrient impairment. (Adapted from USEPA [2017].)

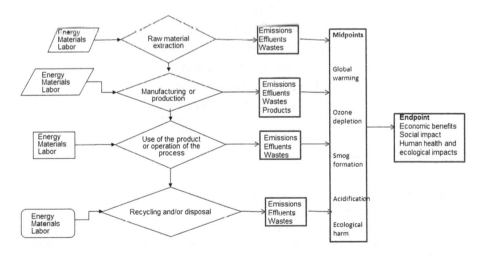

FIGURE 7.6
Environmental, social, and economic impacts over the life cycle of a product or process. (Adapted from Allen and Shonnard [2012].)

considered as a component in global warming or the toxicity of a particular component in water.

Eighteen environmental indicators by the OECD (www.oecd.org/ innovation/green/toolkit) are shown in Figure 7.7 for a sustainable

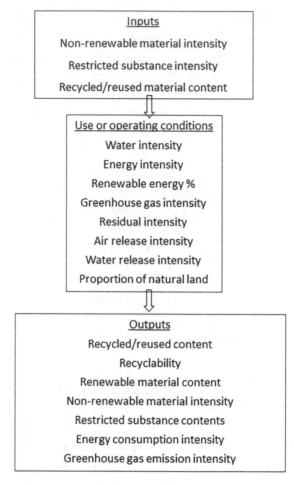

FIGURE 7.7
OECD sustainable manufacturing indicators (OECD 2018).

manufacturing tool kit. The Guide provides advice for collecting local data and calculating indicators to help improve immediate performance, evaluate, and communicate the environmental measures of a product, a facility, or at the organizational level to various stakeholders. Guidance is provided for obtaining the data.

Other entities have also developed indicators. The USDOT (2016) has focused on environmental aspects for its transportation department. The U.S. Army Corp of Engineers has developed planning guidance for sustainability and report progress in a scorecard for GHG reductions, reduction in energy intensity, use of renewable energy, reduction in potable water intensity, in fleet petroleum use, and green buildings (www.usace.army.mil/Missions/Sustainability/Sustainability-Energy-Scorecard/).

7.4 Economic Aspects and Indicators

Monetizing environmental costs and impacts can be quite difficult. Some indicators have been developed to consider some of these impacts as traditional accounting procedures do not take aspects of sustainable engineering into account. Internal costs are those, for example, within a manufacturing plant such as salaries and cost of materials. External costs are related to cost to the society due to the facility, for example, lost days of work due to poor air quality.

Allen and Shonnard (2012) indicated a framework based on the American Institute of Chemical Engineering Center for Waste Reduction Technologies (AIChE CWRT 2000). This included five tiers:

- Tier I: Costs in normal engineering projects such equipment, materials, labor, utilities.
- Tier II: Administrative and regulatory costs such as waste treatment equipment and expenses, permit costs, sampling, emission inventorying, and reporting.
- Tier III: Liability costs such as compliance or remediation obligations, fines, penalties, compensation, damages.
- Tiers IV and V: Costs internal and external, related to improving environmental and societal performance and to insurance, customers, lenders, and communities. These can be difficult to estimate.

Engineers Canada (2016) has indicated that economic analysis should include all aspects of capital, operating, maintenance commissioning, decommissioning, social and environmental costs. Therefore, the entire life cycle should be considered. Mitigating and/or adapting to climate change costs should also be considered. Performance indicators for costs can include environmental full cost account that includes environmental, social, and economic costs and benefits, LCA tools (calculatelca.com), and the System of Environmental-Economic Accounting (SEEA) for environmental costing (http://unstats.un.org/unsd/envaccounting/seea/asp).

7.5 Social Aspects and Indicators

As one of the pillars of sustainability is social, then aspects related to this must be included in any sustainability assessment. Engineers are not usually experts in all aspects and thus should consult with social scientists and other stakeholders to determine impacts and issues of designs and projects on society. The responsibility of engineers is to protect the environment and society as it is part of their code of ethics. Regional, local, and community concerns should be identified to determine potential impacts on traditional and cultural values. First Nations in particular should be consulted to seek input on project acceptance over the life cycle of the project as early as possible. Values need to be defined and incorporated in the project to minimize social impacts.

Blechman et al. (2017) have indicated that stakeholders should work with engineers in all phases of the project. In the early stage, they can help to define needs, concerns, and understand the social and political aspects in the community that can influence the project. Later, they can help to establish various criteria for evaluation of options. Even during operation, they can help to identify concerns and monitor agreements. In sustainable engineering, stakeholder engagement is to enlarge and enhance engineering knowledge to ensure a successful project/process. If necessary, experts can be consulted regarding social and political aspects. These can include public relations experts, media consultants, survey designers, data managers, budget experts, facilitators, and mediators. Depending on the size, complexity, and degree of controversy, different levels of community participation can be expected. These can range from social media, information sessions, and town halls to forming working groups with stakeholder representation.

Different groups have examined what social impacts should be included in sustainability assessments. The United Nations has conceived of various social indicators (UNDSD 2001). However, integration of social sustainability can be difficult for engineers as many of these indicators involve nutrition, poverty, population growth, security, and education. Others directly implicate engineers such as corruption and vulnerability to natural hazards, access to energy and drinking water, and sanitation.

Indicators are divided into themes of poverty, governances, health, education, demographics, and natural hazards and are more focused at the global level than on engineering projects. Kölsch et al. (2008) developed social indicators for product and process design. Although there are some similarities to the UN list, additional indicators are included for employment accidents and diseases, gender equality, research & development spending and professional training.

Corporate social responsibility (CSR) is a tool that is used by companies and is not necessarily appropriate for engineers. Major categories for product manufacturing developed by the UNEP include the worker, consumer, local community, society, and value chain actors. Kelly et al. (2017) had indicated that although they can be adapted for infrastructure projects, they are not for consideration of the entire life cycle and the social LCA (ISO 14040 2006) is more appropriate. The Partnership for Sustainable Communities has devised various indicators according to type, scale, level of urbanization, and issue. Some indicators are shown in Table 7.3.

TABLE 7.3

Indicators for the Partnership for Sustainable Communities (Kelly et al. 2017)

Indicator Topic	Scale	Urbanization Level	Issue
Housing	County	Rural	Access and equity
Land use	Municipality	Suburban	Affordability
Transportation	Neighborhood	Urban	Sense of place and community
	Region		Economic competitiveness
			Environmental quality
			Public health

The GRI is currently used by many organizations and contains several social components in the Sustainability Social Guidelines (GRI 2018). Companies' social responsibilities for their employees are related to security, employment practices, health and safety, and training. ISO 14000 series (particularly ISO 14031) has guidelines on social dimensions.

In Envision, the ISI rating system, previously discussed, includes social equity and justice that are of particular concern. In the Envision checklist, improving community quality of life category includes questions involving the community, well-being, and purpose subsections.

7.6 Available Tools for Sustainable Engineering

The Institute of Chemical Engineers (IChemE) (UK) has an extensive set of indicators that covers all three domains as shown in Table 7.4. IChemE environmental indicators included: resources (energy, material, water, and land) and impacts (acidification, global warming, human health, ozone depletion, photochemical ozone, wastes, and ecological health) (IChemE 2002). Economic indicators include (value added, per unit sales or direct employer, gross margin per direct employee, return on average direct employer, percent change in capital, and R&D expenditures as percentage sales. Social indicators include benefits as percentage of payroll expense, employee turnover, promotion rate and working hours lost as percentage of total hours worked, income and benefit ratio, lost-time accident frequency, expenditure on illness and accident prevention/payroll expense, and number of complaints per unit value added.

Sustainability tools include a wide variety of indicators. However, there are many common aspects among the indicators. Energy, water, and materials input and output flows. The impact of the cost of manufacturing, impact of wastes, emissions, recycling, treatment, health, and ecological impact and land use changes.

TABLE 7.4

IChemE (2002) List of Indicators

Environmental	Economic	Social
Resources	Value added	Benefits as percentage of payroll expense
Energy	Value added per unit value of sales	Employee turnover
Material	Value added per direct employee	Promotion rate
Water	Gross margin per direct employee	Working hours lost as percentage of total hours worked
Land	Return on average capital employee	Income and benefit ratio
Impacts	Percent change in capital employee	Lost time due to accidents
Acidification	R&D expenditures as percentage sales	Expenditure on illness and accident prevention/payroll expense
Global warming		Number of complaints per unit value added
Human health		
Ozone depletion		
Photochemical ozone		
Wastes		

The SITES rating system from the Sustainable Sites Initiative (2018) examines land development for various infrastructure projects. Categories include site context, predesign and assessment, site design, construction, education, and performance monitoring. Silver, gold, and platinum certification levels, similar to LEED are provided. Ecosystem and biodiversity sustenance are encouraged.

Sustainability Tracking, Assessment, and Rating System (STARS) are the sustainability tools for assessing and rating communities. Seven goal areas are covered in the rating system (STAR Communities 2018), including built environment, climate, energy, economy and jobs, education/arts and community, equity and empowerment, health and safety, and natural systems. Several communities have now achieved certification.

Greenroads (Greenroads Foundation 2018) is used for transportation infrastructure such as roadways. Energy, material, and waste reduction credits are awarded. Integration of sustainability practices into design and construction in a project. Bronze, silver, gold, or evergreen certification levels are awarded based on points obtained.

Infrastructure Voluntary Evaluation Sustainability Tool (INVEST) (Federal Highway Administration (FHWA 2018) is also developed for transportation. It is used for self-evaluation to enable planners to integrate sustainability practices into their projects. Performance measurements are determined, then the project is scored, and finally there is reflection on the results and how to improve the measures undertaken.

Many tools also weight the indicators and it is difficult to see the details on this. The weighting should be comparative for the options considered on the same basis. Engaging with stakeholders can help the ranking processes. Local needs and knowledge can be obtained from the stakeholders. In addition, communication is enhanced. Reporting from the rankings can be done in a visual way to enhance communication such as radar plots (Ainger and Fenner 2014).

For the chemical industry, the Institute for Sustainability of the AIChE has developed a Sustainability Index to quantify sustainability aspects. There are seven factors that are used and are publicly available with 5–6 metrics each: strategic commitment, environmental performance, safety performance, product stewardship, social responsibility, innovation for sustainability, and management of the value chain. The data for each factor are weighted and a spider chart is then formed. For each company the data is incorporated into the overall SI. Chin et al. (2015) performed an analysis of ten companies that indicated that they were committed to sustainability and produced a report conforming to the GRI. In general, it seems the overall performance has improved since 2007 using the SI. The environmental and social performance aspects were not improving but this could be due to a lack of quantifiable data.

The Gauging Reaction Effectiveness for the ENvironmental Sustainability of Chemistries with a Multi-Objective Process Evaluator (GREENSCOPE) is a sustainability assessment tool used to evaluate equipment and in the design of chemical processes by the U.S. EPA Office of Research and Development. A total of 140 indicators are listed on material efficiency, energy, environment and economy. Figure 7.8 shows how it can be used to improve a process.

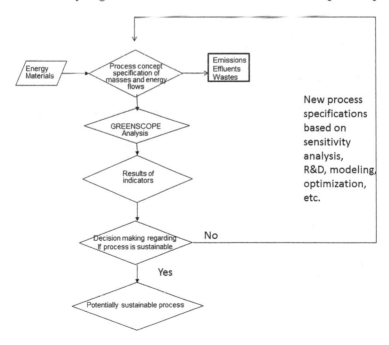

FIGURE 7.8
Applicability of tools for process improvement. (Adapted from Ruiz-Mercado et al. [2016].)

The GREENSCOPE tool allows for quantifying process sustainability and LCI generation. The indicators include information related to process performance, feedstocks, utilities, equipment, and outputs for a part of a process or a complete one. Different processes can then be compared to determine if a raw material or process has an improved sustainability performance. In addition, the designer or the researcher can implement this methodology to evaluate the sustainability performance after making process modifications and at the end generate a manufacturing-stage LCI for LCA applications.

The WAR, GREENSCOPE, and Sustain Pro have been used together for an ammonia production case study (Ruiz-Mercado et al. 2016). Sustain Pro (Carvalho et al. 2013) is useful for comparing design alternatives by screening and evaluating alternatives based on indicators. It involves data collection, flow sheet decomposition, indicator calculations, an ISA algorithm, a sensitivity analysis, and generation of alternatives. The overall integration of the three tools is shown in Figure 7.9. The Waste Reduction Algorithm (WAR) algorithm (Young et al. 2000) for conceptual design can be obtained online. It performs an assessment using the eight environmental impact categories. In the case study, there was an indication by WAR that the ammonia product itself could cause major environmental impact instead of the emissions. GREENSCOPE (Ruiz-Mercado et al. 2012) was then able to show more specifically, where impacts could be. Sustain Pro was then used to determine alternatives for reduced consumption of heat or improving the final product

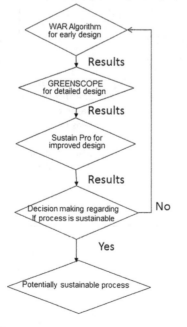

New process specifications
Based on sensitivity analysis,
R&D, modeling, optimization, etc.

FIGURE 7.9
Integration tools for process design. (Adapted from Ruiz-Mercado et al. [2016].)

recovery. The solutions to this complex problem were not simple but are useful for engineers.

Ruiz-Mercado et al. (2013) performed a sustainability assessment using sustainability indicators for biodiesel production. They used the design tool GREENSCOPE. This tool was developed to enable designers and decision makers the ability to incorporate sustainability in chemical processes and improve the process over the life cycle. The tool models chemical processes and assessing sustainability using indicators and is based on the WAR which has eight environmental impact indices for the environmental evaluation. Figure 7.9 shows where GREENSCOPE has a direct influence. The biodiesel production process was evaluated to show where process improvements could take place and improve the sustainability of the process. The environmental indicators showed an overall good performance. The data is presented as a radar graph for environmental, energy, and economic aspects with 0% as the worst case and 100% as the best. However, some aspects could be improved such as finding a solution for a waste oil produce by recycling and finding a way to avoid excess methanol use. The energy indicators showed overall good energy intensity. Output streams in some cases are at high temperatures. These may be overdesigned or there are opportunities for changing the operating conditions of the units. For the economic indicators, there was a strong dependence on the feedstock price. It made up 58% of the total manufacturing cost. There is uncertainty in feedstock and sales prices, which was not taken into account. Investment costs could be recovered in about 3 years after start-up, which is good. Treatment costs for water and waste were also good. Therefore, this tool is useful for identifying areas for improvement of a chemical process.

7.7 Examples of Indicator and Tool Use

Romero et al. (2017) presented an application of indicators for improvement of a pressurized water system in Callosa d'En Sarrià (Alicante, Spain) for drinking and irrigation purposes. The indicators were sociocultural, economic, and environmental. The EPANET hydraulic software was used. For the case study, energy savings were 12.3%, cost savings were 15.5%, the energy footprint of water was reduced by 15.0%, and GHG were reduced by 12.3%. In addition, the distributed water volume increased by 9.1%. Replacement of a pressure reduction valve by a pump working as a turbine produced 103,700 kWh/year of renewable energy. The methodology employed is presented in Figure 7.10. Various energy indicators were used. Environmental indicators included GHG emissions. Social indicators included development of new jobs to reduce migratory movement. All results are summarized in Table 7.5. The sustainability indicators were based on numerical simulation

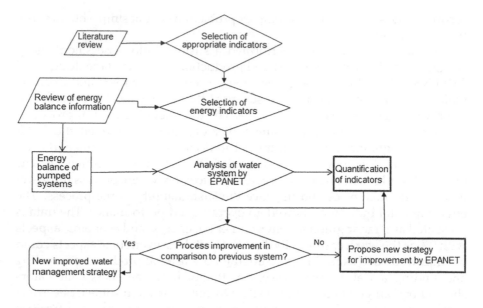

FIGURE 7.10

Process of sustainability improvement using EPANET. (Adapted from Romero et al. [2017].)

TABLE 7.5

Indicators and Improvements Obtained (Romero et al. 2017)

Indicator	Improvement
Emissions CO_2 equivalent (kg/year)	12.3%
Energy Indices	
Network energy efficiency	2.56
Excess of supplied energy	2.49
Friction dissipated energy by the network	15.8
Annual consumed energy (kWh/year)	12.3
Energy footprint of water, energy per unit volume (kWh/m³)	19.6
Energy efficiency of pumps	3.2
Economic	
Total energy cost	15.5%

and management tools and allowed improvement in the sustainability and security of the water systems. Elements to recover or to reduce the consumed energy, improved the energy efficiency. Regulation of the pumps and hydraulic machines improved the energy efficiency and reduced GHG emissions.

Another example of the use of indicators was performed by Chalise (2014) to enable assessment of the sustainability of the wastewater treatment plant (WWTP) systems and to improve performance. The qualitative and quantitative indicators are applied for industrial wastewater treatment but can also be suitable for municipal wastewater treatment.

Four dimensions of sustainability: environment, society, economy, and technology were employed. The methodology of the development of the indicator and weighting scheme consisted of:

- Review of the literature of indicators for assessment of the sustainability of WWTPs
- Selection of appropriate indicators regarding the options and the local conditions
- Preparation of a list of indicators under various themes
- Consultation with stakeholders with regard to indicator selection and scoring

Relative scoring schemes consisted of a 0–100 score for the lowest to the highest performing options. Absolute scoring schemes were based on a fixed scale adopted from various organizations (UNEP, WHO, etc.). Tables 7.6 and 7.7 shows selected indicators and their scoring schemes.

Indicators of the environmental dimension of impacts of the system and their references include site footprint, short- and long-term impact upon habitat, land use and diversity, and environmental management track record. Economic sustainability is a very important element of sustainable development. It allows making changes to enhance sustainability and economic growth at reduced environmental impact.

Traditional economic indicators focus on economic impacts on the community while evaluating the financial performance of the options.

TABLE 7.6

Four Categories of Sustainability Indicators for Wastewater Treatment Evaluation (Chalise 2014)

Economic	Environmental	Social	Technical
NPV (total project cost)	Air quality energy	Hazardous	Complexity
Capital costs	consumption	materials	Durability
Operation and maintenance costs	Environmental	Local job	Flexibility
Decommissioning costs	toxicity	creation and	Reliability
Ease of obtaining necessary permits	GHG emissions	diversity	Upgradability
User costs	Impacts on	Management	
Financial uncertainty	biodiversity	practices	
Financial recoveries	Hazardous output	Public health	
Land footprint	Liquid waste	and safety	
Logistics	discharge	Worker health	
NPV (net present value of total project cost)	Reuse, recycle	and safety	
Operation and maintenance costs	Water use		
User costs			

TABLE 7.7

Scoring Schemes of Selected Indicators (Chalise 2014)

Environmental Indicators	Scoring Scheme						
	0	25	33	50	66	75	100
Quantity of liquid/solid waste discharge	**Regressive, Linear Relationship**						
Quality of Liquid/Solid Waste Discharge	Option meets regulatory requirements only		Discharge quality surpasses regulatory requirements		Discharge quality significantly surpasses regulatory requirements		Discharge meets industry best-practice standards and guidelines
ReUse							
GHG	**Positive, Linear Relationship** **GHG Calculator**						
Environment Management Track Record	Company has no environmental management system (EMS) but track record of meeting regulatory requirements.	Company has basic EMS and track record of meeting regulatory requirements		Company has basic EMS and track record of implementing sustainable environmental practices above regulatory requirement		Company has detailed EMS and track record of implementing sustainable environmental practices above regulatory requirements	Compliance with ISO14001-2004 and ISO14004-2004 standards, and to meet industry best-practice guidelines

(Continued)

TABLE 7.7 (Continued)

Scoring Schemes of Selected Indicators (Chalise 2014)

Social Indicators	Scoring Scheme			
	0	33	66	100
Public/Worker Health and Safety	Option meets all applicable public health and safety laws and regulations, and involves technologies or methods for which no health and safety regulations exist	Option meets all applicable public health and safety laws and regulations and involves no unregulated health and safety concerns	Option surpasses regulatory requirements	Meets industry best-practice standards and guidelines
Hazardous Material	0	33	100	
	Use of materials assigned the signal word "danger"	Use of materials assigned the signal word "warning"	No use of materials with an assigned signal word	
Economic Advantages for the Local Community	0	25	50	75
	No portion of goods and/or services will be provided by local businesses	0%–25% of goods and/or services will be provided by local businesses	25%–50% of goods and/or services will be provided by local businesses	50%–75% of goods and/or services will be provided by local businesses

(Continued)

TABLE 7.7 (*Continued*)

Scoring Schemes of Selected Indicators (Chalise 2014)

Social Indicators	Scoring Scheme			
	0	25	50	75
Local Job Creation & Diversity	Negligible impact on employment opportunities for locals	Significant (>25% or more of the labor component of the total budget) temporary or seasonal employment opportunities for locals	Significant, temporary, or seasonal employment opportunities for locals, including deliberate efforts to hire minority and/or low-income groups	Significant, permanent employment opportunities for locals
Community support for the option/project	Large negative public sentiment against the option, e.g., group complaints, local media reports/campaign	Small negative public sentiment against the option, e.g., Individual complaints	No identifiable public sentiment	Small positive public sentiment in favor of the option, e.g., individual submissions of support
Nuisance	High nuisance level	Medium nuisance level, long duration (average >1 week/month)	Medium nuisance level, med duration (average 1 day –1 week/month)	Medium nuisance, short duration (average <1 day/month)

(Continued)

TABLE 7.7 (Continued)

Scoring Schemes of Selected Indicators (Chalise 2014)

Social Indicators			Scoring Scheme		
Management	Company has no CSR policy	Company has basic CSR policy	Company has basic CSR policy and track record of implementing social practices above regulatory requirements	Company has detailed CSR policy and track record of implementing social practices above regulatory requirements
Economic Indicators			**Scoring Scheme**		
NPV			**NPV Calculator**		
Financial Recoveries	0	25	50	75	100
	No potential for financial recovery	Potential recovery of 0%–10% of total cost	Potential recovery of 10%–20% of total cost	Potential recovery of 20%–30% of total cost	Potential recovery of over 30% of total cost
Land Footprint			**Regressive, Linear Relationship**		
Ease of obtaining necessary permits	0	25	50	75	100
	Difficulties expected in obtaining many permits	Difficulties expected in obtaining a small number of permits	Permit approval expected to be conditional for a small number of permits	Approval process expected to be straightforward	

(Continued)

TABLE 7.7 (Continued)

Scoring Schemes of Selected Indicators (Chalise 2014)

Economic Indicators	Scoring Scheme				
	0	**25**	**50**	**75**	**100**
Interference with activities on site	Major delays (>1 day)	Moderate (1 h–1 day) >1/week	Moderate (1 h–1 day) <1/week	Minor (<1 h) >1/week	No anticipated impact
Logistics	More complicated logistical requirements. Many previously unused suppliers **(0)**	More complicated logistical requirements. Some previously unused suppliers **(33)**	Simple logistical requirements. Some previously unused suppliers **(66)**	Simple logistical requirements. Option uses previous logistic supply chain **(100)**	

Technical Indicators	Scoring Scheme				
	0	**33**	**66**	**100**	
Technical Performance Reliability	Unproven reliability to achieve design criteria – pilot testing required/underway	Proven reliability to achieve design criteria for other applications. For previous unproven durability reliability – pilot testing completed	Proven reliability to achieve design criteria in similar applications/environment	Direct past experience proving reliability at a satisfactory level for similar applications	

(Continued)

TABLE 7.7 (Continued)

Scoring Schemes of Selected Indicators (Chalise 2014)

Technical Indicators	Scoring Scheme			
	0	33	66	100
Flexibility/ Robustness	System cannot process elevated loadings and flows above design criteria	System can process small elevated loadings and flows above design criteria	System can process medium elevated loadings and flows above design criteria	System can process large elevated loadings and flows above design criteria
Technical Uncertainty	New technology. No track record. No experience with pilot testing	New technology, pilot testing in site conditions completed. Existing technology without previous application to these project conditions	Technology in broad industrial use. No previous experience (directly) using	Previous experience with use/ implementation of the technologies
Design complexity	Consultant, new or bespoke design	Consultant, standard design (greater than 240 h)	Standard design, consultant (under 240 h)	Package plant – minor design required (under 40 h)

Source: Adapted from Chalise (2014).

The social dimension represents the welfare of society and if the population is affected by facilities and procedures. Potential impacts should be transparent and communicated to the stakeholders. Decisions must be made with public input. Incorporation of the indicators into a multi-criteria analysis framework enables decision-making for mitigating life cycle costs, regulatory risks, energy and GHG emissions, enhancing reuse opportunities, and social acceptability. Selection and weighting of the evaluation criteria must fit the specific local conditions. Evaluations for new and improved wastewater treatment technologies can be developed to enhance sustainability, reliability, and flexibility to benefit users, the community, and the environment.

The technical dimension enables comparison of the technical aspects of the various options to provide the best service for the community while being that is simple to apply with minimum complexity and technical difficulties. The required technology should be easily available with proven successful application in similar regions and environmental constraints. It should be able to meet the relevant regulatory requirements. The scoring can be done based on the past performance of the technology and provisions of the option to cope with any service interruptions, such as an emergency.

Although ideally, sustainable practices should be incorporated at an early design stage, they can be done at any stage. As many WWTPs have been operating for many years, these are excellent candidates for process optimization, to enhance performance sustainability.

Various kinds of equipment, pumps, and motors consume resources and energy, have workers safety concerns, and lead to environmental impacts. Those impacts need to be considered as a part of wastewater treatment evaluation, design, and implementation. Industry needs to develop evaluation criteria and matrices that can be used in decision-making and operation.

7.8 Conclusions

Environmental, economic, and social aspects must be evaluated to improve process/product/system sustainability. Engineers have the most influence at this level. Selection of indicators is required to do a sustainability assessment. The indicators should cover all three sustainability aspects. Many organizations have suggested various sets of indicators and guidance for selecting the indicators. Once the appropriate indicators are chosen, then data must be collected and an assessment is performed, often using a multi-criteria analysis framework. The indicators can assist in evaluating process improvements, selection of materials, minimization of environmental, social, and economic impacts, and selection of process alternatives to improve process sustainability.

References

3M, 2018. Sustainability Report. www.3m.com/3M/en_US/sustainability-report/. Accessed June 25, 2018.

AIChE CWRT (American Institute of Chemical Engineers' Centre for Waste Reduction Technologies), 2000. *Total Cost Assessment Methodology*, New York: AIChE.

Ainger, C. and R. Fenner, 2014. *Sustainable Infrastructure: Principles into Practice*, London: Thomas Telford Ltd.

Allen, D.T. and D.R. Shonnard, 2012. *Sustainable Engineering, Concepts, Design and Case Studies*, Upper Saddle River, NJ: Pearson Education, Prentice Hall.

Association for Canadian Engineering Companies (ACEC), 2016. Sustainable Development for Canadian Consulting Engineers with new 2016 preface. www.acec.ca/publications_media/acec_publications/sustainability/index. html. Accessed June 25, 2018.

ASTM, 2015. Standard Guide for Evaluation of Environmental Aspects of Sustainability of Manufacturing Processes, E2986-15, West Conshohocken, PA: ASTM International. www.astm.org.

Beloff, B., J. Schwarz and E. Beaver, 2001. Use of sustainability metrics to guide decision making. *CEP*, 2002: 58.

Blechman, F., C. Davidson and E. Kelly, 2017. Community participation. In: *Engineering for Sustainable Communities*, W.E. Kelly, B. Luke and R.N. Wright (eds), Roston, VA: ASCE Press, pp. 179–200.

Carvalho, A., H.A. Matos, and R. Gani, 2013. SustainPro—A tool for systematic process analysis, generation and evaluation of sustainable design alternatives. *Computers and Chemical Engineering*, 50: 8–27.

Chalise, A.R., 2014. Selection of Sustainability Indicators for Wastewater Treatment Technologies, MASc thesis, Concordia University, Montreal, Canada.

Chin, K., D. Schuster, D. Tanzil, B. Beloff and C. Cobb, 2015. Sustainability trends in the chemical industry. *CEP*, 111(1): 36–40.

Engineers Canada, 2016. National Guideline on Sustainable Development and Environmental Stewardship for Professional Engineers, September 2016, https://engineerscanada.ca/publications/national-guideline-on-sustainable-development-and-environmental-stewardship. Accessed June 25, 2018.

EPA, 2016. Waste Reduction Algorithm: Chemical Process Simulation for Waste Reduction. www.epa.gov/chemical-research/waste-reduction-algorithm-chemical-process-simulation-waste-reduction.

Fiksel, J., 2009. *Design for the Environment. A Guide to Sustainable Product Development*. 2nd edition, New York: McGraw Hill.

Fiksel, J., T. Eason and H. Frederickson, 2012. A Framework for Sustainability Indicators at EPA. Cincinnati, OH: National Risk Management Research Laboratory, EPA Number: EPA/600/R/12/687 for additional details on the framework and taxonomy.

Fiksel, J., 2015. *Resilient by Design*. Washington, DC: Island Press.

FHWA (Federal Highway Administration), 2018. INVEST-version 1.3. www.sustainablehighways.org/. Accessed June 25, 2018.

Greenroads Foundation, 2018. Greenroads Rating Systems. www.sustainablehighways.org/. Accessed June 25, 2018.

GRI, 2018. GRI Standards Resource Download Center, Amsterdam, The Netherlands. www.globalreporting.org/standards/resource-download-center/. Accessed June 25, 2018.

IChemE, 2002. Sustainable development progress metrics recommended for use in the process industries. nbis.org/nbisresources/metrics/triple_bottom_line_indicators_process_industries.pdf. Accessed December 5, 2018.

ISO, 2006. Environmental Management-Life Cycle Assessment-Principles and Framework. www.iso.org.

Kelly, W., K. Reddy, G. Lovegrove, S. Fisher, L. Lemay, C. Davidson and B. McDowell, 2017. Social aspects. In: *Engineering for Sustainable Communities*, W.E. Kelly, B. Luke, R.N. Wright (eds), Reston, VA: ASCE Press, pp. 99–112.

Kölsch, D., P. Saling, A. Kicherer, A. Grosse-Sommer and I. Schmidt 2008 How to measure social impacts? A socio-eco-efficiency analysis by the SEEBALANCE method. *International Journal of. Sustainable Development*, 11(1): 1–23.

NRC, 2011. *Sustainability and the U.S. EPA*. Washington, DC: The National Academies Press. ISBN 10: 0-309-21252-9.

OECD, 2018. OECD Sustainable Manufacturing Indicators. www.oecd.org/innovation/green/toolkit/oecdsustainablemanufacturingindicators.htm. Accessed June 25, 2018.

Pinter, L., 2004. Compendium of Sustainable Development Indicator Initiatives. www.iisd.org/library/compendium-sustainable-development-indicator-initiatives. Accessed June 25, 2018.

Romero, L., M. Pérez-Sánchez and P. Amparo López-Jiménez, 2017. Improvement of sustainability indicators when traditional water management changes: A case study in Alicante. *AIMS Environmental Science*, 4(3): 502–522.

Ruiz-Mercado, G.J., R.L. Smith and M.A. Gonzalez, 2012. Sustainability indictors for chemical processes: I. Taxonomy. *Industrial and Engineering Chemistry Research*, 51: 2309–2328.

Ruiz-Mercado, G.J., M.A. Gonzalez and R.L. Smith, 2013. Sustainability indicators for chemical processes: III Biodiesel case study. *Industrial and Engineering Chemistry Research*, 52: 6747–6760.

Ruiz-Mercado, G.J., A. Carvalho and H. Cabezas, 2016. Using green chemistry and engineering principles to design, assess, and retrofit chemical processes for sustainability. *ACS Sustainable Chemical Engineering*, 4(11), 6208–6221.

Shell Global, 2017. Sustainability Reports. www.shell.com/sustainability/sustainability-reporting-and-performance-data/sustainability-reports.html. Accessed June 25, 2018.

Sikdar, S.K., 2003. Sustainable development and sustainability metrics. *AIChE Journal*, 49(8): 1928–1932.

Sikdar, S.K., D. Sengupta and R. Mukherjee, 2017. *Measuring Progress towards Sustainability: A Treatise for Engineers*. Cham, Switzerland: Springer International Publishing.

STAR Communities, 2018. STAR Community Rating System, Version 2.0. www.starcommunities.org/. Accessed June 25, 2018.

Sustainable Sites Initiative, 2018. SITES Rating System. www.sustainablesites.org/. Accessed June 25, 2018.

USDOT United States of America, Department of Transportation, 2016. Strategic Sustainability Performance Plan, Submitted June 30, 2016, Office of the Secretary of Transportation | Office of Sustainability and Safety Management. www.transportation.gov/sustainability. Accessed June 28, 2018.

UNDSD (United Nations Division of Sustainable Development), 2001. *Indicators of Sustainable Development: Guidelines and Methodologies*. New York: UNDSD.

United Nations, 2018. SDG indicators Global Database beta 0.2.43, Statistics Division. https://unstats.un.org/sdgs/indicators/database/. Accessed June 28, 2018.

USEPA, 2012. Framework for Sustainability Indicators at EPA. EPA/600/R/12/687. October 2012, Office of Research and Development, National Risk Management Research Laboratory, Sustainable Technology Division.

USEPA, 2017. EPA's Report on the Environment, Indicators A–Z. https://cfpub.epa.gov/roe/indicators.cfm. Accessed June 25, 2018.

Young, D.M., R. Scharp and H. Carbezas, 2000. The waste reduction (WAR) algorithm: environmental impacts, energy consumption and engineering economics. *Waste Management*, 20: 605–615.

8

Implementation of Sustainable Engineering Practices

8.1 Introduction

Sustainability has gained significant attention in the past decades, particularly by engineers. The definition of sustainability by the Brundtland Commission in the 1970s, however, is quite vague and difficult to implement. More recently, tools have been developed to assist in integrating sustainability into design, particularly for buildings and infrastructure. None of these has been adapted universally. In the 1970s, environmental considerations became important. Now, aspects such as resource conservation, societal acceptability, energy minimization, use of renewable energies, and mitigation of climate change, among others, must be included in project/process/system considerations. This chapter will provide examples of sustainable engineering practices and challenges and needs for the future education and research.

8.2 Integration of Sustainability Concepts into Engineering Practices

According to the Association of Consulting Engineering Companies (ACEC)—Canada report on consulting engineering and sustainability (Boyd 2016), conservation, preservation, prediction, and consultation are four main aspects to be included in engineering design. Conservation includes minimization of water, energy, and materials by recycling or use of renewable energies. Durability and site restoration after decommissioning must all be included in the planning. Environmental laws must be followed for biodiversity preservation and health and safety. Human rights must be respected and climate change impacts must be considered. Cultural considerations are also becoming more important. Prediction involves designing for resilience in a changing environment mainly due to climate change. Affected communities must be consulted, as they are important stakeholders in projects.

Climate change is another very important challenge for sustainability. There are two main approaches, mitigation and adaptation. Mitigation involves the reduction in carbon dioxide emissions, in particular. This can be related to reducing energy use, using cleaner fuels, and possible carbon capture. This applies to transportation, buildings, industry, and environmental remediation such as waste management. Adaptation involves new design for projects, as historical data will not apply. Greater uncertainty and thus large factors of safety will be needed.

Clients are becoming more interested in sustainability. Many companies are now participating and producing sustainability reports using the Global Reporting Initiative (GRI) as seen in Chapter 6. Many government agencies also have committed to sustainability. Public participation has been mainly during the environmental assessment phase of a project. This is usually late. They need to be included from the concept of the projects to uncover issues and to communicate progress to the stakeholders.

Procurement also needs to be included in sustainability. This includes no sweatshop labor or unfair labor practices or corruption, use of fair trade products, and support of local development. Engineers must improve sustainability of projects and processes for the future. Determination of indicators at an early stage of the project in consultation with stakeholders is essential. The engineer has the most influence in the early stage of the project. However, there are opportunities during commissioning, operation, and decommissioning to improve sustainability. Climate change is also causing more uncertainty in designs. New tools can help. Traditionally, projects have to meet the client and regulator's requirements. The society is adding new requirements for the project in terms of sustainability. Engineers have an ethical requirement to face these new requirements. New tools are being developed to assist in the assessment but none is perfect. For example, LEED and BREEAM are restricted to buildings.

Innovation is a key aspect for sustainable engineering. These projects are riskier and require more time initially. However, payoffs will be in the future with reduced operating costs. According to the World Federation of Engineering Organizations (WFEO 2015), engineers can contribute to sustainable practices in many ways. Some of these include harvesting renewable resources in ways to ensure continuous supply, minimization of nonrenewable resources, processing resources with little to no wastes, designing and building of infrastructure and processes with minimal waste and environmental impact throughout the life cycle, and development of clean renewable energy sources. Human needs must be met for ensuring adequate living and health standards. In other words, sustainable engineering involves the use of natural resources in a cost-effective way for the support of the human and natural environments. The approach should be as close to a closed loop as possible as proposed in the circular economy. In Chapter 5, various sustainable practices for resources, environmental restoration, and energy production and use were discussed. The next section will demonstrate some case studies of sustainable engineering practices.

8.3 Examples of Sustainable Practices

8.3.1 Industrial Park in Haiti

In Haiti, the construction of the Caracol Industrial Park was to take place. This was to be one of the most important infrastructure projects in the country. When it came to the production of a drinking and industrial water production facility, the mandate for the turnkey project was removed. At this moment, the engineering consulting company SNC-Lavalin took the initiative to design the plant.

Initially, a designer with little experience in working in developing countries, proposed a reverse osmosis system to remove components from the groundwater such as manganese. However, this option was very energy intensive, particularly for a country that gets 95% of their energy from diesel or fuel oil. In addition, the maintenance and operation of the system would be very difficult. The proposed solution was an ion exchange system. The resins would eliminate water hardness and other elements such as manganese that gives color, bad taste, and toxicity, if the water is consumed over a long period of time.

The advantages of the system included:

- The process consumed 98% less pumping energy than reverse osmosis that requires high pressures. The process did not include any mechanical parts such as motors or gears that would have to be replaced due to use.
- The process included the local community in the village of Caracol that already had a salt production system from sea water (Figure 8.1) that was necessary for the recharging of the anionic resins; in other words, the village found a new market for their salt for the system that would require 4 metric tons/day for the production of the water.

When the plant was designed, all equipment was selected so that it could be mounted or moved by an overhead crane of 5 metric tons. In other words, the bridge was able to accommodate the 19 pumps and 10 reservoirs for the softening as well as the 1 metric ton bag of salt for the plant and move them directly by a truck of van without the need of a crane or other equipment. The project was started in 2011 and inaugurated in 2012.

FIGURE 8.1
Production of salt in the village of Caracol (satellite view).

The project consisted of:

- A 400 m² building
- Water treatment for drinking (6,000 m³/day)
- Water for industry (6,000 m³/day)
- Water for firefighting (2,160 m³/h)

The system consisted of a phase 1 water treatment reservoir (drinking) of 500 m³ and one for industrial water of 1,500 m³. The disinfection was by chlorination (available locally with two peristaltic pumps for 24 h use). The cost was US $3,000,000 for the drinking water production. The design period was 50 h for the preliminary design and 200 h for the detailed design.

8.3.1.1 Environmental Aspects

The process conformed to the rules and regulations of the World Development Bank. An environmental assessment, environmental management plan, and management of greenhouse gases (GHGs) were all according to the norms of the World Development Bank.

The technology used 98% less electricity from heating oil generators, 2–5 MWh of energy for 500 m³/h of water. This reduced the production of GHGs. The pumps have their own frequency regulator that reduced their energy at start-up and optimized their continuous operation. Translucent panels were used for the panels of the building, thus reducing lighting requirements.

Ventilation and temperature control were ensured by 10 cm (4 in.) vertical spaces between the panels. All wash water was recuperated and sent to the treatment. None was sent to the stormwater drainage system. The water from the softening systems was high in hardness. This was beneficial for the industrial plant that required high alkalinity for its coagulation/flocculation process. A landfill according to the norms of the World Development Bank was built and all buildings had their own septic system along with infiltration fields.

The process did not produce any dangerous wastes. Disinfection was by chlorination. The peristaltic pumps were able to pass ten times higher concentrations of chlorine than regular dosage pumps and thus lower volumes of stored solution and less handling were needed. They did not produce noise.

8.3.1.2 Social Aspects

Salt was purchased from the local community at a rate of 4 metric tons/day. A social impact study was performed according to the norms of the World Development Bank. Six offices of the six neighboring communities were organized to manage the hiring of personnel and favored hiring from the local region. Twenty percent of the construction workers were female, which was a first in Haiti, where normally 0.5% is female. Efforts were also made for deploying social responsibility, conserving cultural sites, avoiding negative social impacts, and supporting the neighboring communities as performed by the World Development Bank and client.

8.3.1.3 Economic Aspects

The plant was to employ 3–5 local people as technicians and chemists. The salt production needed up to 50 employees. Others were needed to transport the salt. Therefore, all created employment opportunities for the local community. In addition, the reduction in energy will save energy costs.

8.3.1.4 Ethics

The project was designed with the concept of sustainability. The building was well lit and ventilated, local salt was used in the production, a crane was used to move the equipment, and the project produced essential water for drinking, industry, and firefighting.

8.3.2 Case Study of the London 2012 Olympics

The London 2012 Olympics was proposed to ensure sustainability in the design, building, operation, decommissioning, and reinstatement of infrastructure (ODA 2013). Sustainability was integrated into the design, procurement and contract management, planning, building, and operation phase along the lines of biodiversity, energy, environmental impact, materials, water, and waste.

Strategy development was the first phase. The vision can be seen in Figure 8.2 along the lines that leave no trace (no environmental damage), zero waste to landfill (reducing waste and maximizing reuse/recycling), zero harm (no accidents, injury, or incidents), and leave a positive legacy (environmental, social, and economic).

The infrastructure was quite unique in that it was temporary so there were no other examples to base the procedures on. Table 8.1 shows the priority targets and outcomes. Success factors were identified such as securing committed leadership, embedding sustainability in procurement and contracts, establishing, monitoring and reporting key performance indicators, and reviewing all audit processes such as environmental management plans.

To deliver the strategic vision and targets, four areas were identified. These included sustainable design, procurement and contract management, town planning, and building and operations.

Sustainable design procedures and results included:

- Establishing a working group for design
- Developing guidance for technical specifications
- Working with suppliers to review products and materials

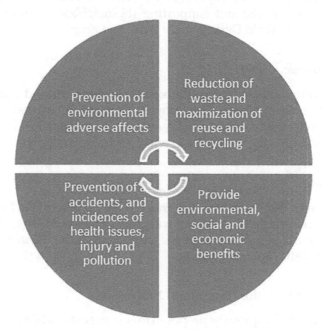

FIGURE 8.2
Strategy and vision for sustainability for London 2012. (Adapted from Aukett [2012].)

TABLE 8.1

Priority Targets in Sustainability Planning for the London 2012 Olympic and Paraolympic Games

Objective	Final Outcome
Reduction of carbon emissions and minimization of carbon footprint through hiring 85% of commodities to reduce embodied carbon	86% reached
Reduction of venue footprint by 25% to reduce need for materials, air conditioning, lighting, etc.	47% reached
Reduction of heating, ventilation, and air conditioning (HVAC) cooling by 70% and maximization of natural cooling to reduce energy, carbon and GHG emissions	82% reached
Minimization of particulate matter by 80% to improve air quality	90% reached at 47 generator sites
Purchase of at least 20% of materials that are recycled or from a secondary target to reduce material and carbon impacts	Could not be determined
Reuse and recycling of 90% of materials from installation and deconstruction of facilities to reduce wastes	99% reached

Source: Adapted from Aukett (2012).

- Reducing need for extra ventilation and cooling
- Designing standard modules to reduce waste

Procurement and contract management steps consisted of:

- Introducing sustainability ambitions to suppliers in the expression of interest phase
- Pre-ITT (invitation to tender), initial discussion on sustainability considerations at meeting
- ITT phase 10 sustainability questions required responses
- ITT review, Sustainability Procurement Check Sheet completed
- Contract award, sustainability credential needed by suppliers.
- Contract development, sustainability policies, and procedures embedded in the contracts
- Contract management framework and audit tool used to track sustainability

Communications and negotiations were essential to address challenges, particularly of technical versus sustainability, and time issues. Sustainability requirements were often overwhelming for suppliers. Lack of information on the suppliers' own products required training.

Town planning required environmental impact, protection, and management plans to be implemented. Environmental Impact Assessment

(EIA), surveys, and impact assessments had to be performed in consultation with regulators, stakeholders, and planners. Keeping track was essential, all had to be done for a short-term event that presented special challenges. Regular meetings were required. Many surveys were available for land already.

The process for managing impacts and risk at the operations phase included management of environmental risks (air and water quality, noise, ecological preservation), management of waste and resources for reduction, reuse and recycling, and contractors were required to operate sustainably. Stakeholders engagement was ensured at different levels (close engagement, consultation, and information, or information only). Compliance auditing was undertaken on a regular basis; issues and incidents were managed and reporting from all suppliers was required.

Some of the overall conclusions and accomplishments included:

- Clear vision and approach on technical information, such as presence of carbon footprint, are needed.
- Strategic approach needed to be developed for all aspects.
- Budget and resource needed to be addressed early.
- Targets needed to be prioritized and reviewed.
- Design teams and suppliers should be engaged and challenged to improve performance.
- Support is needed for many suppliers and many now have new skills in sustainable practices.
- These lessons although taken for a unique event can be used for many projects.

Various microreports are also available online providing more detail on the sustainability of various aspects of the Olympic Games. For example, an evaluation (ODA 2011) was performed by the design teams to determine which type of pavement would be more sustainable. Asphalt, poured concrete, permeable asphalt, block paving, and porous gravel were some of the options considered. The unit for comparison was one square meter of paving.

The criteria for comparison based on the Building Research Establishment Green Guide Specification included:

- Embodied carbon dioxide
- Recycled content
- Recyclability at end of life
- Weight of materials to reduce handling risks
- Avoided carbon dioxide by recycling

- Urban heat island effect
- Pavement depth
- Olympic Delivery Authority (ODA) green guide specification rating

Other aspects included aesthetics, buildability, cost, wet weather performance, and vehicle loading. Conclusions were reached that despite the high-embodied carbon dioxide content, asphalt was more sustainable due to the requirement for less material. Less depths were needed that reduced the material requirement. The tool CEEQUAL was useful for optimizing material use and energy. Increasing the content of recycled material increased sustainability. It provided a central location for information for discussion by stakeholders and enabled maximization of sustainability using several elements.

Under a contract from the ODA, all the civil engineering, landscaping, and public realm works at the Olympic Park were assessed and verified using CEEQUAL in 17 separate package assessments, all achieving 'Excellent' rated awards. These individual assessments included the Enabling Works (North and South Areas), the Landscaping and Public Realm in the South and North Parks, the District Heating and Cooling Network, Overbridges and Stadium Bridges, and the Primary Foul Sewer and Pumping Station. The scores from the individual assessments were aggregated on a construction value-weighted basis, giving an overall weighted 'Excellent' CEEQUAL score of 93.8%. Furthermore, two of the 17 Olympic Park projects achieved very high scores of 98.3% for the CEEQUAL Assessment: Olympic Park North Park Structures, Bridges and Highways (SBH Lot 1) and Olympic Park Wetland Area Bridges.

8.3.2.1 Economic Activity

Overall, in creating the Park, 2.8 million U.K. construction professionals were involved and over 4,000 long-term jobs were created for the new technology, design, and research center. They generated due to the 4 million visitors during the summer and another 800,000 visitors per year will visit the swimming center. A total of 8,000 new homes, 12 new schools and nurseries, 3 health centers, and a new library were planned after the Games. Two million metric tons of heavily contaminated earth was remediated in addition to the regeneration of the River Lea through the Park. More than 6.5 km of waterways will be monitored in the park. Efforts toward enhancing wildlife habitats incorporated bat roosts, frog ponds, kingfisher walls, otter holts, and planting of wild flowers. Features such as LED lighting, photocell switches, efficient fixtures, and the irrigation system fittings were employed for reduction of energy requirements. Renewable energy production onto the grid included wind turbines and photovoltaic cells for the lighting columns. Overall, 98% of Olympic Park demolition work materials were reclaimed for reuse and recycling, and greater than 650 bird and bat boxes were installed across the Olympic Park (ODA 2013).

London 2012 was the first Summer Olympic and Paralympic Games to measure the carbon footprint over the entire project (CEEQUAL 2013). It showed that by using an independent sustainability assessment tool like CEEQUAL planning, executing of projects and maintenance work during operation can be influenced to enhance sustainability practices.

8.3.3 Sustainable Remediation Using GOLDSET

Various tools can be used such as carbon footprint, ecological footprint, energy efficiency, and life cycle assessment (LCA) but these cover only the environmental aspects. Some multicriteria analytical ranking or scoring systems such as GOLDSET (Golder Associates 2018) are available that cover all three aspects. An example of this is provided as follows:

The GOLDSET-CN approach includes project description and determination of alternatives. Selection of the appropriate indicators, evaluation of the options by data entry and selecting scoring and weighting and finally interpreting the results and recommendations as shown in Figure 8.3. The process can be done across the project life cycle as shown in Figure 8.4.

FIGURE 8.3
Five-step process using GOLDSET (Golder Associates 2018).

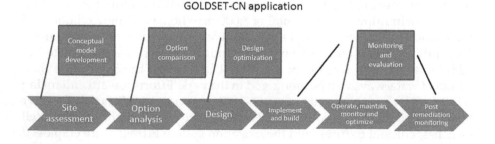

FIGURE 8.4
Application of GOLDSET-CN across the project life cycle (Golder Associates 2018).

A case study was undertaken on the derailment of a train in a peat bog in the province of Quebec in Canada. About 280,000 L of petroleum, which was mainly diesel, spilled and impacted 12,000 m³ of peat and soil in a ca. 7,000 m² area that was 2.4 m in depth. An emergency pumping of the fuel from the water was performed, the train cars were removed, and a confinement trench was installed. The bog was ecologically sensitive and the local community considered it to be an ecological reserve. Three options were evaluated: natural attenuation, partial excavation with risk management, and complete excavation. Indicators were then selected. Estimates of energy and emissions of GHG were calculated in the GOLDSET module as shown in Table 8.2. Scoring and weighting were then performed, as shown in Table 8.3.

Stakeholder involvement was a key part of this sustainable remediation framework. University research leaders as well as members of the regulating

TABLE 8.2

Sample Indicators and Benchmarks Set for the Case Study

Indicator	Unit	Period 6	Period 7	Planned at Completion	Accepted Variation (%)
Soil quality	m³	0	0	600	10
GHG emissions	total CO_2 eq.	0.18	0.54	47.17	25
Energy consumption	GJ PFE	2.68	8.22	647.7	25
Hazardous waste	kg	0	400	5,730	50
Impacts on biodiversity	—	45	45	90	
Energy consumption per cubic meter excavated soil	GJ PFE/m³	0	0	0.83	15
GHG emissions per cubic meter excavated soil	Total CO_2 eq./ m³	0	0	0.06	15

TABLE 8.3

Sample Indicators Scoring and Weighting (Golder Associates 2018)

Economic Indicator	Natural Attenuation	Partial Excavation and Risk Management	Total Excavation
Remediation Option			
Net present value	0	100	100
Potential litigation	90	90	100
Financial recoveries	75	50	25
Environmental reserve	100	100	100
Economic advantages for local community	0	50	50
Technological uncertainty	0	50	100
Logistics	100	50	0

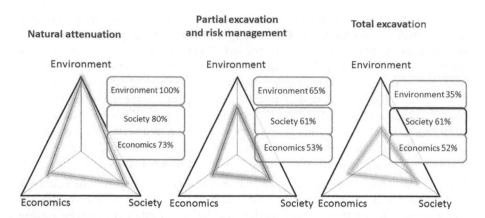

FIGURE 8.5
Comparison of remediation options using Goldset-CN for derailment in a peat bog. The most balanced shadowed triangle indicates the most sustainable option.

agencies were invited to contribute to weigh in on the issues at hand, and provide pertinent comments and questions. These key stakeholders contributed to the composition and procedural design of expert reviews, which helped ensure that all stakeholders found the results of this process credible.

The remediation scheme chosen involved partial excavation and *in situ* remediation. Despite the sustainability of natural attenuation, the regulators did not accept this option. Special floating walkways were installed to access the pilot plots to minimize impact on the vegetation. Solar panels were used to reduce GHG emissions for the blower. Truck-mounted equipment was replaced by tripod mounted or manual augers to reduce vegetation damage. The local university provided expertise to restore the peat bog. Excavation and transport were reduced which in turn reduced 250 metric tons of carbon dioxide emissions. Local landowners were kept informed which enabled participation and acceptance.

GOLDSET provided a transparent communication tool (results shown in Figure 8.5), and enabled the evaluation and tracking of sustainability indicators over the life cycle of the project. Environmental, social, and economic performance can be enhanced by this process.

8.4 Future Needs

8.4.1 Role of Education

Education programs have introduced the sustainable development concepts in to courses to undergraduate engineering students. An industrial survey in 2013 indicated that energy use and efficiency, recycling and reuse, life cycle analysis and corporate social responsibility are key educational needs (Fergus et al. 2013). Special courses have been developed and sustainability has been integrated into course material. Specialization in sustainable development in new degrees may also be another option. Integration of sustainable practices in courses has been difficult as some lecturers are opposed and others are not trained how to do them. Environmental aspects are more easily integrated than social and equity issues. Real-world cases (Steiner and Posch 2006) are required to demonstrate the interdisciplinarity and transdisciplinarity in a complex problem-solving environment. This approach provides a better, dynamic, and more sustainable learning environment.

Sustainability has also become a requirement of engineering program accreditation. In the United States, Accreditation Board for Engineering and Technology (ABET) indicates that students must be prepared for aspects of professional practice that includes sustainability among other aspects. Graduate attribute mapping where learning outcomes are determined and tracked is becoming more popular internationally. Guidance to integrate sustainability in course curriculum is still lacking. Sustainability goes beyond the environmental aspects currently in many courses. In Canada also, Engineers Canada, the national organization of the 12 engineering regulators that licenses Canadian members of the profession, expects graduates to have various skills. This includes understanding of the interactions and the uncertainty of engineering on the economic, social, health, safety, legal, and cultural aspects of society. Some organizations such as Engineers without Borders USA (www.ewb-usa.org) exist in various countries (almost 50) and involve professionals and students for executing projects in developing countries to ensure that the water, sanitation, energy, and infrastructure needs for the communities are met.

Education is key to training future engineers in sustainability. This trend toward interdisciplinarity in engineering education is reflected by an increasing number of interdisciplinary sustainability initiatives at universities and research institutions (CAGS 2012; De Graaff and Ravesteijn 2001). In Canada, a prime example is the University of Victoria that has been the recipient of funding for its *Training Program in Interdisciplinary Climate Science* (University of Victoria 2017). The University of Toronto has applied an interdisciplinary approach to identifying the role of engineers in solving complex global problems—including those related to sustainable development—at the *Centre for Global Engineering* (CGEN) (University of Toronto 2009). In its *Cinbiose* interdisciplinary environmental research centre, *L'université du Québec à Montréal* (UQAM) has collaborated with the World Health Organization since 1998 on problems in several fields such as health care, ecosystem dynamics, urban ecosystem governance, and climate change (UQAM 2017). Western University offers graduate-level courses on interdisciplinary approaches to sustainability studies, and also maintains a standing research faculty contributing to sustainability research across 33 different academic disciplines (Western University 2017).

In the United States of America, initiatives in multidisciplinary environmental research are well established. Stanford University's *Precourt Energy Efficiency* Center (Stanford University 2016) is an example. Universities by setting curriculum standards have the moral responsibility to educate their graduates to play a crucial role in developing a socially just, ecologically aware, and economically responsible society. At the same time, engineers have the obligation to develop and implement design, construct, and manage techniques that minimize environmental and energy footprints. In addition, engineers must also be able to work in multidisciplinary teams that incorporate perspectives from public policy, economics, and social responsibility. These demands place a unique burden on engineering educators to design programs that will train engineers for future challenges.

Engineers must be able to work in multidisciplinary teams incorporating public policy, economics, and social responsibility (Figure 8.6). In light of the above, Concordia University has established the Concordia Institute for Water, Energy, and Sustainability Engineering (CIWESS) that provides a unique interdisciplinary training in water, energy, and sustainability engineering (concordia.ca/ciwess). The specific objectives of this training program are: to catalyze through collaboration, internships, enhanced research opportunities in sustainability; to train highly qualified personnel in an interdisciplinary manner for public, parapublic, and industrial sectors; to maintain and enhance interdisciplinary areas of teaching and research; and to attract external research funding and foster relationships with external researchers and internal Concordia researchers with similar interests.

The training program (Mulligan 2017) is producing trainees with unique knowledge and skills related to sustainable water and energy systems through a combination of multiple programmatic pathways, such as undergraduate minors, graduate degrees and courses, capstone courses, research seminars, internships, conferences, and public outreach.

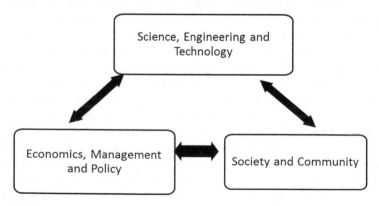

FIGURE 8.6
Interdisciplinary aspects of the training program.

Since interdisciplinarity is an essential prerequisite of any type of research related to sustainability, this program incorporates it at the level of content and program. The underlying philosophy of interdisciplinarity of the program is revealed in the schematic representation in Figure 8.7. All research content while being grounded in scientific and technological aspects nevertheless incorporates economic/policy and society/community aspects. A key mechanism to facilitate interdisciplinarity in research content is through the constitution of supervisory committees that will include members from all aspects.

Since environmental and social issues are by their very nature complex and interrelated, students are required to cross disciplinary boundaries in order to collaborate with those in other disciplines. Interdisciplinary collaboration requires skills that demand modifying traditional ways of thinking and being open to novel means of cross-disciplinary communication. Internships are provided to allow trainees to work in the modern collaborative workplace to improve their academic and professional strengths

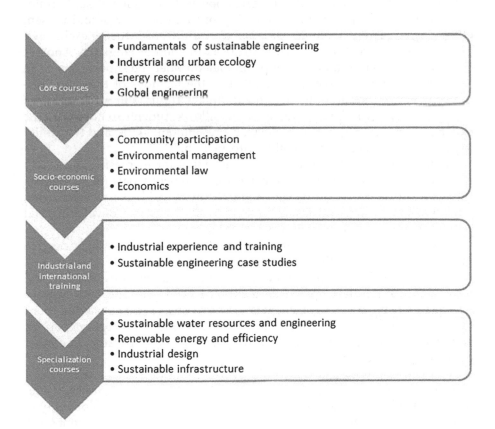

FIGURE 8.7
Components of sustainable engineering graduate programs.

and acquire program-relevant work experience. They gain important soft skills (communications, teamwork, interpersonal cooperation) essential in today's work environment, learn more about the expectations and needs of employers, develop independence and maturity, and take advantage of net-working opportunities. The skills obtained enable students to work in policy development, governmental agencies, international organizations, industry, and nongovernmental organizations (NGOs). An internship is a required element of every trainee's program. An annual research event showcasing the achievements of the program provides a venue for trainees to present and discuss their work.

The program facilitates these collaborations by integrating opportunities that arise from these research exchanges into training. In addition, by enhancing international exposure these collaborations will foster globally sustainable work practices in trainees. For example, trainees develop international expertise through these collaborations in water-related disciplines as water engineering and management, land use planning, or water environmental studies.

Trainees are required to attend workshops and professional skills training. These workshops enhance the ability of trainees to write and present effectively, to plan and manage projects, to study abroad during exchanges, to understand ethical practices, and to speak in another language. Another integral component is an internship completed under the supervision of an experienced engineer/computer scientist in the facilities of the participating company. In addition, the university allows trainees to work full time for 4 months at the industry. This structure allows students to maximize the experience of working in teams, develop presentation skills, and the ability to prepare written reports and other information.

Training and education programs are essential for increasing the ability of engineers to address the Sustainable Development Goals (SDG) targets. Innovative initiatives across educational institutes should be shared. Industry, government agencies, and academia need to work together developing and promoting innovation and new ways of thinking.

Informing the public is also an essential element that universities need to include, particularly in their research programs. For example, in a research project in CIWESS, residents involved in lake associations are consulted and information of the development of technologies to maintain and improve lake quality to enhance engagement is provided on a regular basis. Solutions for waste management also must involve the public to incorporate, adopt, and maintain new technologies. Training programs for operators and other personnel are essential.

In a workshop hosted at Concordia University (Montreal, Canada June 9, 2016) with Polar Knowledge on waste management and waste to energy solutions for northern communities, there were discussions of the feasibility of different waste to energy (W2E) solutions for communities of different sizes. Community size was an important factor. Namely, the solutions for waste management and W2E are different for small communities [<500 residents] compared to medium [500–2,000 residents] and large communities [>2,000 residents] in the North. For medium and large communities, landfilling best practices, compaction, composting, incineration or gasification with heat recovery, or refuse-derived fuel (RDF) boiler. For small communities, options include landfilling best practices, compaction, and incineration with heat recovery.

Key aspects for waste management solutions include education, training, buy-in from the community, and to keep trained the community members in the community. Availability of parts and maintenance can be major challenges in remote communities. Community engagement with project planning/design is essential.

8.4.2 Role of Engineering Organizations

Various engineering organizations have formed sustainable development committees to inform their members regarding sustainable engineering and have formed guidance documents. Some of these are by the American Society of Civil Engineers (ASCE) and the U.K. Institute of Civil Engineers (ICE) and the Canadian Society for Civil Engineering (CSCE). The Committee on Engineering and the Environment enables WFEO and the global engineering profession to support the achievement of the UN Millennium Development Goals through the development, application, promotion, and communication of:

- Environmentally sustainable engineering practices and technologies
- Adaptation of infrastructures to the impacts of a changing climate
- Assessment and promotion of clean technologies and engineering practices to mitigate climate change
- The engineering perspectives on the international elements of the agricultural supply chain to United Nations agencies and commissions, national members of the Federation, and other international NGOs
- Development of guidelines for practicing engineers on responsible environmental stewardship and sustainable practices in various areas of practice including mining

The WFEO represents more than 20 million engineers with extensive expertise to contribute to the achievement of the UN Sustainable Development 2030 Goals (WFEO 2015). Engineers must work with organizations, other experts, governmental organizations, and the communities to develop technologies, policies, and frameworks. Engineering societies in many disciplines must also work together toward sustainable engineering. Sharing of case studies can assist in promoting and understanding sustainable practices and implementation. The WFEO Model Code of Practice for Sustainable Development and Environmental Stewardship (WFEO 2013) was developed and adopted at the September 2013 General Assembly.

The ten principles of the Code of Practice include:

1. Maintaining and continuously improving awareness and understanding of environmental stewardship, sustainability principles, and issues related to the field of practice.

2. If knowledge is not adequate to address environmental and sustainability issues, consult with others with the required expertise.

3. Global, regional, and local societal values should be incorporated to include local and community concerns, quality of life, and other social concerns related to environmental impact. Traditional and cultural values must be included.

4. Sustainability aspects should be incorporated at the earliest possible stage and employ applicable standards and criteria.

5. Costs and benefits of environmental protection, ecosystem components, climate change and extreme events, and sustainability should be incorporated in the economic viability of the work.

6. Environmental stewardship and sustainability planning should be over the life cycle for planning and management activities and efficient, sustainable solutions should be employed.

7. A balance between environmental, social, and economic factors should be achieved to contribute to healthy surroundings in the built and natural environment.

8. An open, timely, and transparent engagement process for both external and internal stakeholders should solicit input in and respond to economic, social, and environmental concerns.

9. Regulatory and legal requirements must be met and exceeded by applying best available, economically viable technologies and procedures.

10. In the cases where there are threats of serious or irreversible damage and scientific uncertainty, risk mitigation measures should be implemented in a timely fashion to minimize environmental degradation.

Engineers of all disciplines are involved in the development of technologies to improve life but must assure that impacts (social, economic, and environmental) are minimized. For example, Johnston (2016) indicated that process, civil, mechanical, electrical, instrumentation, and control engineers should work together with economists and scientists to ensure sustainable wastewater treatment and management solutions. Conferences such as STS Forum in Japan (held annually), Engineering for Sustainability (held in Denver, CO, February 18–19, 2016), and the 2015 Engineering Triennial Summit of the ICE/ASCE/

CSCE held in London, United Kingdom, are excellent examples of efforts to establishing collaborations between various societies. Climate change is also creating challenges. Engineering is essential in achieving the UN Sustainability Goals. Figure 8.8 shows some engineering actions for each of the selected goals according to the WFEO. Rating systems such as Envision or CEEQUAL can assist in the assessment of infrastructure project through goals and indicators and are a step in the right direction. Continual development of these and other tools is needed for other types of projects and processes.

An example of Envision ratings is the Platinum Award given to the New Champlain Bridge Corridor project in Montreal, Canada (Infrastructure Canada 2018). The bridge was designed to last 125 years. Some of the features of the project included:

- Design of the stormwater drainage system to take climate change into consideration
- Offset of GHG emissions by carbon credits
- Use of LED lights to reduce light pollution effects on migratory birds and energy requirements
- Recycling of 45% of the construction waste onsite and of the other 54% that was recycled, only 1% was sent to landfill
- Marshlands restoration and creation of new spawning habitats to offset loss of fish habitats, wetlands, and bird sanctuaries during construction of the bridge

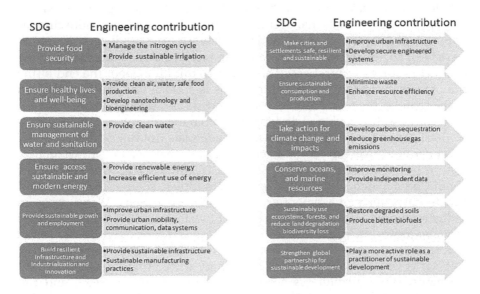

FIGURE 8.8
WFEO contributions toward selected UN Sustainable Development goals (WFEO 2015).

LCA can be complicated and difficult to implement. A life cycle approach, however, is essential. Life cycle sustainability assessment approaches are being developed which may be more appropriate than LCA due to inclusion of the social and economic aspects. Policies and incentives are needed to promote incorporation of sustainability in projects. Industries have made efforts to reduce wastes, emissions, energy uses, and report their progress through GRI standards.

8.4.3 Innovation for Sustainable Engineering

Bugliarello (2010) has identified various challenges for innovation for engineering, in the areas of materials, water and energy, IT, and bioengineering. Resources conservation and waste management are particularly relevant. Standards and regulations should not restrict innovation needed to advance sustainable practices. Nanotechnology, composite materials (especially with natural materials will improve material function and resource protection. The linkages of water and energy need to be addressed by considering both together not separately as is the current practice. Therefore, engineers and professional societies need position papers, metrics, and methodologies for guiding decisions, case studies, and data for enabling new ways of thinking in devising more effective water and wastewater treatment systems, reducing leakage for distribution systems, collection of energy from solar and wind power systems, and replacement of hydrocarbon fuels.

Materials that can be reused or recycled are highly desirable. Reuse and recycling of metals such as metals are essential. This includes aluminum, cobalt, copper, nickel, rare earth and platinum, iridium, gold and silver. Linear practices of waste management should be replaced by a more circular approach of disassembly, reuse, and recyclability. New products will need to be developed, in addition to new cans, car, airplane, and bicycle parts, rebar, etc. that is currently practiced. This requires thinking about the end of the life of a product from the beginning. This is a major challenge in addition to the complexity of the products. Metal sorting and separation are also part of the process. Research and development from the laboratory to pilot industrial scale are essential and supported by adequate funding.

Fenner and Ainger (2016) have indicated that water infrastructure should be monitored for water use and treatment. Harmful substances should be avoided and natural systems adopted such as wetlands and ponds for water management and restoring ecosystems. Water reuse in industry should be encouraged instead of water wasting. Indicators such as water footprinting can assist in promoting water resource conservation and smart water use.

Water is very connected to human activity with a high social and political nature; its availability is not equally distributed worldwide and can vary significantly seasonally and annually. Water use by industry can create significant public protests particularly when there is scarcity of water.

Water management is thus highly important and can drive innovation. As was recently addressed at the Civil Engineering Triennial Summit in London, climate change must be addressed now due to the increased risk of poverty, social inequity, terrorism, and conflict (Perks 2015).

The Innovation for Cool Earth Forum (2017) formulated a statement from the steering committee to further innovations toward net-zero emission of carbon dioxide in light of the Paris Agreement at the COP21. The statement is based on the forum of over 100 experts from many fields from 80 countries.

Some of the key points include:

- A peak of carbon dioxide emissions must be in the near future.
- Technological innovation is essential through zero emission technologies (renewables and nuclear energy), low carbon products, low carbon infrastructures, international cooperation, and financial and social innovation.
- Industrial sector participation in innovation and its diffusion.
- Utilization of a systems approach for diffusion of market-ready technologies (wind and solar power).

Perks et al. (2017) have provided some suggestions on what civil engineers can do. These include making sure existing infrastructure is working in an optimal manner. Leakage, for example, in water systems is a major problem due to water and energy wastage. Low cost, easy to operate, and socially acceptable solutions also must be identified instead of going with the usual practice. Consultation with stakeholders with "fresh eyes" through all stages of the project planning can bring new alternatives for consideration. Carbon neutral approaches for new infrastructures should be adopted. For example, as water pumping is a major user of energy, the use of gravity should be optimized. Future designs must be economically efficient to be affordable for the public and reduce poverty and must reduce and avoid energy consumption as much as possible to mitigate climate change. Key performance indicators are needed to measure progress.

8.5 Conclusions

There are presently many challenges for engineers. Education programs need to be expanded for engineers to include sustainable practices not as a special course but integrated into existing courses. Improved tools are needed. Envision and other tools are a good start but they are complex, hard to address over the life cycle of a project or process and many aspects are lacking, particularly regarding social equity and cultural issues. More partnerships are needed between social scientists, physical science, health and engineering professionals, education, governance and society. In summary, technology, education, regulation, and standards are all essential to promote and implement sustainable engineering practices. Interdisciplinary programs are necessary for training engineers in the circular economy.

Engineers need to work together and be more involved in decision-making at all stages of the project. They should become more involved in local or regional activities to assist in the decision-making. They need to consult with stakeholders for input regarding concerns and to adapt to local conditions. Even during the construction and/or operation phases, engineering should be able to address concerns and provide advice on the sustainability of a project. Research is needed to develop innovative solutions to this changing world under the influence of climate change and increasing uncertainty, deteriorating infrastructure, introduction of new chemicals into the environment, centralization and lock-in of technologies, and growing population to name a few. Engineers have an ethical requirement to rise to this challenge. They have indicated that they want to be involved in sustainable engineering practices. The role of engineers in sustainable development has been undervalued but it is critical. However, action is needed now.

References

Aukett, A., 2012. Delivering Sustainability for Games Venues and Infrastructure, Learning Legacy, Lessons Learned from the London 2012 Games Construction Project, Olympic Delivery Authority December 2012, LOC2012/SUS/CS/0015. http://london2012.com/learninglegacy. Accessed September 30, 2017.

Boyd, J., 2016. Association Sustainable Development for Canadian Consulting Engineers. With new 2016 preface. Association of Consulting Engineering Companies. Canada, Ottawa, ON. www.acec.ca/publications_media/acec_publications/sustainability/index.html.

Bugliarello, G., 2010. Emerging and future areas of engineering, Engineering: Issues Challenges and Opportunities for Development Produced in conjunction with: World Federation of Engineering Organizations (WFEO) International Council of Academies of Engineering and Technological Sciences (CAETS), International Federation of Consulting Engineers (FIDIC), UNESCO Publishing, Paris, France, pp. 56–59.

CAGS, 2012. Graduate Student Professional Development: A Survey with Recommendations, Canadian Associate for Graduate Studies: Professional Skills Development for Graduate Students. www.cags.ca/documents/publications/working/Report%20on%20Graduate%20Student%20Professional%20Development%20%20-%20A%20survey%20with%20recommendations%20FINAL%20Eng.OCT%202012.pdf. Accessed January 30, 2017.

CEEQUAL, 2013. Eric Hughes Award for Outstanding Contribution to Improving Sustainability in Civil Engineering Awarded to London 2012 Olympic Park. www.ceequal.com/news/eric-hughes-award-for-outstanding-contribution-to-improving-sustainability-in-civil-engineering-awarded-to london-2012-olympic-park/. Accessed July 7, 2018.

De Graaff, E. and W. Ravesteijn, 2001. Training complete engineers: Global enterprise and engineering education. *European Journal of Engineering Education*, 26: 419–427.

Fenner, R. and C. Ainger, 2016. Challenges and opportunities. In: *Sustainable Water*, C. Ainger and R. Fenner (eds), London: ICE Publishing, pp. 21–31.

Fergus, J.W., C. Twigge-Molecey and J. McGuffin-Cawley, 2013. Sustainability in materials education. *JOM*, 65: 935–938.

Golder Associates, 2018. GOLDSET. https://golder.goldset.com/portal/Default.aspx. Accessed July 7, 2018.

ICEF, 2017. 4th Annual Meeting, Tokyo, Japan. www.icef-forum.org/pdf2018/pastevent/icef2017-report-e.pdf. Accessed October 5, 2017.

Infrastructure Canada, 2018. New Champlain Bridge Corridor Project Receives ENVISION™ Platinum Award. www.newswire.ca/news-releases/new-champlain-bridge-corridor-project-receives-envision-platinum-award-684599881.html. Accessed July 31, 2018.

Johnston, A., 2016. Wastewater management. In: *Sustainable Water*, C. Ainger and R. Fenner, (eds), London: ICE Publishing, pp. 123–146.

Mulligan, C.N., 2017. An Interdisciplinary Research and Training Program in Sustainability-CIWESS, Engineering Solutions for Sustainability 3. Toward a Circular Economy. February 18–19, 2017, Denver, CO.

ODA, 2011. Assessing the Sustainability of Pavement Design Solutions, Learning Legacy, Lessons Learned from the London 2012 Games Construction Project, Olympic Delivery Authority October 2011, ODA 2010/374. http://london2012.com/learninglegacy. Accessed October 31, 2017.

ODA, 2013. Sustainability. http://learninglegacy.independent.gov.uk/themes/sustainability/index.php.

Perks, A., 2015. *A Bridge Too Far-Civil Engineering in Transition, 2015 Engineering Triennial Summit*, London: ICE/ASCE/CSCE.

Perks, A., G. Lovegrove, E. Tam, A. Khan and C. Mulligan, 2017. Rising above routine practice. In: *International Conference on Sustainable Infrastructure, Sustainable Cities for an Uncertain World*, New York, October 26–28, 2017.

Stanford University, 2016. Precourt Energy Efficiency Center (PEEC) Stanford University. https://peec.stanford.edu/. Accessed January 30, 2017.

Steiner, G. and A. Posch, 2006. Higher education for sustainability by means of transdisciplinary case studies: An innovative approach for solving complex real-world problems. *Journal of Cleaner Production*, 13: 877–890.

University of Toronto, 2009. New Centre for Global Engineering (CGEN). http://news.engineering.utoronto.ca/new-centre-global-engineering-cgen/. Accessed January 30, 2017.

University of Victoria, 2017. NSERC CREATE Training Program in Interdisciplinary Climate Science. http://climate.uvic.ca/UVicCREATE/. Accessed January 30, 2017.

UQAM, 2017. Cinboise. https://cinbiose.uqam.ca/. Accessed January 30, 2017.

Western University, 2017. Centre for Environment and Sustainability. http://uwo.ca/enviro/about_us/index.html. Accessed January 30, 2017.

WFEO, 2013. WFEO Model Code of Practice for Sustainable Development and Environmental Stewardship Think Global and Act Local. www.wfeo. org/wp-content/uploads/code-of-practice/WFEOModelCodePractice_ SusDevEnvStewardship_One_Page_Publication_Draft_en_oct_2013-3.pdf. Accessed July 6, 2018.
WFEO, 2015. WFEO Engineers for a Sustainable Post 2015. World Federation of Engineering Organization, UN Scientific and Technological Communities Major Group, July 22, 2015, Version 1.6.

Appendix A: UN Sustainable Development Goals for 2030

Selected from www.un.org/sustainabledevelopment/sustainable-development-goals/ (accessed August 8, 2018).

Goal 6

Facts

- At least 1.8 billion people globally use a source of drinking water that is fecally contaminated.
- Between 1990 and 2015, the proportion of the global population using an improved drinking water source has increased from 76% to 91%.
- But water scarcity affects more than 40% of the global population and is projected to rise. Over 1.7 billion people are currently living in river basins where water use exceeds recharge.
- A total of 2.4 billion people lack access to basic sanitation services, such as toilets or latrines.
- More than 80% of wastewater resulting from human activities is discharged into rivers or sea without any pollution removal.
- Each day, nearly 1,000 children die due to preventable water and sanitation-related diarrheal diseases.
- Hydropower is the most important and widely used renewable source of energy and as of 2011, represented 16% of total electricity production worldwide.
- Approximately 70% of all water abstracted from rivers, lakes, and aquifers is used for irrigation.

Goals

- By 2030, achieve universal and equitable access to safe and affordable drinking water for all.
- By 2030, achieve access to adequate and equitable sanitation and hygiene for all and end open defecation, paying special attention to the needs of women and girls and those in vulnerable situations.

- By 2030, improve water quality by reducing pollution, eliminating dumping and minimizing release of hazardous chemicals and materials, halving the proportion of untreated wastewater and substantially increasing recycling and safe reuse globally.
- By 2030, substantially increase water use efficiency across all sectors and ensure sustainable withdrawals and supply of freshwater to address water scarcity and substantially reduce the number of people suffering from water scarcity.
- By 2030, implement integrated water resources management at all levels, including through transboundary cooperation as appropriate.
- By 2020, protect and restore water-related ecosystems, including mountains, forests, wetlands, rivers, aquifers, and lakes.
- By 2030, expand international cooperation and capacity-building support to developing countries in water- and sanitation-related activities and programs, including water harvesting, desalination, water efficiency, wastewater treatment, recycling, and reuse technologies.
- Support and strengthen the participation of local communities in improving water and sanitation management.

___

Goal 7

Facts

- Energy is the dominant contributor to climate change, accounting for around 60% of total global greenhouse gas emissions.

Goals

- Reducing the carbon intensity of energy is a key objective in long-term climate goals.
- By 2030, ensure universal access to affordable, reliable, and modern energy services.
- By 2030, increase substantially the share of renewable energy in the global energy mix.
- By 2030, double the global rate of improvement in energy efficiency.
- By 2030, enhance international cooperation to facilitate access to clean energy research and technology, including renewable energy, energy efficiency, and advanced and cleaner fossil-fuel technology, and promote investment in energy infrastructure and clean energy technology.

- By 2030, expand infrastructure and upgrade technology for supplying modern and sustainable energy services for all in developing countries, in particular least developed countries, small island developing states, and landlocked developing countries, in accordance with their respective programs of support.

Goal 9

Facts

- Manufacturing is an important employer, accounting for around 470 million jobs worldwide in 2009—or around 16% of the world's workforce of 2.9 billion. In 2013, it is estimated that there were more than half a billion jobs in manufacturing.
- Industrialization's job multiplication effect has a positive impact on society. Every one job in manufacturing creates 2.2 jobs in other sectors.
- Small- and medium-sized enterprises that engage in industrial processing and manufacturing are the most critical for the early stages of industrialization and are typically the largest job creators. They make up over 90% of business worldwide and account for between 50% and 60% of employment.
- In countries where data are available, the number of people employed in renewable energy sectors is presently around 2.3 million. Given the present gaps in information, this is no doubt a very conservative figure. Because of strong rising interest in energy alternatives, the possible total employment for renewables by 2030 is 20 million jobs.

Goals

- Develop quality, reliable, sustainable and resilient infrastructure, including regional and transborder infrastructure, to support economic development and human well-being, with a focus on affordable and equitable access for all.
- Promote inclusive and sustainable industrialization and, by 2030, significantly raise industry's share of employment and gross domestic product, in line with national circumstances, and double its share in least developed countries.
- By 2030, upgrade infrastructure and retrofit industries to make them sustainable, with increased resource-use efficiency and greater adoption of clean and environmentally sound technologies and

industrial processes, with all countries taking action in accordance with their respective capabilities.

- Enhance scientific research, upgrade the technological capabilities of industrial sectors in all countries, in particular developing countries, including, by 2030, encouraging innovation and substantially increasing the number of research and development workers per 1 million people, and public and private research and development spending.

Goal 11

Facts

- Half of humanity—3.5 billion people—lives in cities today
- By 2030, almost 60% of the world's population will live in urban areas
- About 95% of urban expansion in the next decades will take place in developing world
- A total of 828 million people live in slums today and the number keeps rising
- The world's cities occupy just 3% of the earth's land, but account for 60%–80% of energy consumption and 75% of carbon emissions
- Rapid urbanization is exerting pressure on fresh water supplies, sewage, the living environment, and public health
- But the high density of cities can bring efficiency gains and technological innovation while reducing resource and energy consumption

Goals

11.6

- By 2030, reduce the adverse per capita environmental impact of cities, including pay in special attention to air quality, municipal and other waste management.
- Should the global population reach 9.6 billion by 2050, the equivalent of almost three planets could be required to provide the natural resources needed to sustain current lifestyles.

Water

- Less than 3% of the world's water is fresh (drinkable), of which 2.5% is frozen in the Antarctica, Arctic, and glaciers. Humanity must, therefore, rely on 0.5% for all of man's ecosystem and fresh water needs.
- Man is polluting water faster than nature can recycle and purify water in rivers and lakes.
- More than 1 billion people still do not have access to fresh water.
- Excessive use of water contributes to the global water stress.
- Water is free from nature but the infrastructure needed to deliver it is expensive.

Energy

- Despite technological advances that have promoted energy efficiency gains, energy use in OECD countries will continue to grow another 35% by 2020. Commercial and residential energy use is the second most rapidly growing area of global energy use after transport.
- In 2002, the motor vehicle stock in OECD countries was 550 million vehicles (75% of which were personal cars). A 32% increase in vehicle ownership is expected by 2020. At the same time, motor vehicle kilometers are projected to increase by 40% and global air travel is projected to triple in the same period.
- Households consume 29% of global energy and consequently contribute to 21% of resultant CO_2 emissions.
- One-fifth of the world's final energy consumption in 2013 was from renewables.

Food

Facts

- While substantial environmental impacts from food occur in the production phase (agriculture, food processing), households influence these impacts through their dietary choices and habits. This consequently affects the environment through food-related energy consumption and waste generation.
- About 1.3 billion metric tons of food is wasted every year while almost 1 billion people go undernourished and another 1 billion are hungry.
- Overconsumption of food is detrimental to our health and the environment.

- About 2 billion people globally are overweight or obese.
- Land degradation, declining soil fertility, unsustainable water use, overfishing, and marine environment degradation are all lessening the ability of the natural resource base to supply food.
- The food sector accounts for around 30% of the world's total energy consumption and accounts for around 22% of total greenhouse gas emissions.

Goals

- Implement the 10-year framework of programs on sustainable consumption and production, all countries taking action, with developed countries taking the lead, taking into account the development and capabilities of developing countries.
- By 2030, achieve the sustainable management and efficient use of natural resources.
- By 2030, halve per capita global food waste at the retail and consumer levels and reduce food losses along production and supply chains, including postharvest losses.
- By 2020, achieve the environmentally sound management of chemicals and all wastes throughout their life cycle, in accordance with agreed international frameworks, and significantly reduce their release to air, water, and soil in order to minimize their adverse impacts on human health and the environment.
- By 2030, substantially reduce waste generation through prevention, reduction, recycling, and reuse.
- Encourage companies, especially large and transnational companies, to adopt sustainable practices and to integrate sustainability information into their reporting cycle.
- Promote public procurement practices that are sustainable, in accordance with national policies and priorities.
- By 2030, ensure that people everywhere have the relevant information and awareness for sustainable development and lifestyles in harmony with nature.
- Support developing countries to strengthen their scientific and technological capacity to move toward more sustainable patterns of consumption and production.
- Develop and implement tools to monitor sustainable development impacts for sustainable tourism that creates jobs and promotes local culture and products.
- Rationalize inefficient fossil fuel subsidies that encourage wasteful consumption by removing market distortions, in accordance with

national circumstances, by restructuring taxation and phasing out those harmful subsidies, where they exist, to reflect their environmental impacts, taking fully into account the specific needs and conditions of developing countries, and minimizing the possible adverse impacts on their development in a manner that protects the poor and the affected communities.

Goal 14

Facts

- As much as 40% of the world oceans are heavily affected by human activities, including pollution, depleted fisheries, and loss of coastal habitats.

Goals

- By 2025, prevent and significantly reduce marine pollution of all kinds, in particular from land-based activities, including marine debris and nutrient pollution.
- By 2020, sustainably manage and protect marine and coastal ecosystems to avoid significant adverse impacts, by strengthening their resilience, and taking action for their restoration in order to achieve healthy and productive oceans.

Goal 15

Facts

- About 2.6 billion people depend directly on agriculture, but 52% of the land used for agriculture is moderately or severely affected by soil degradation.
- As of 2008, land degradation affected 1.5 billion people globally.
- Arable land loss is estimated at 30–35 times the historical rate.
- Due to drought and desertification each year 12 million ha are lost (23 ha/min), where 20 million metric tons of grain could have been grown.
- About 74% of the poor are directly affected by land degradation globally.

- Of the 8,300 animal breeds known, 8% are extinct and 22% are at risk of extinction.

Goals

- By 2020, ensure the conservation, restoration, and sustainable use of terrestrial and inland freshwater ecosystems and their services, particularly in forests, wetlands, mountains, and drylands, in line with obligations under international agreements.
- By 2030, combat desertification, restore degraded land and soil, including land affected by desertification, drought, and floods, and strive to achieve a land degradation-neutral world.

Index

Milton Keynes UK
Ingram Content Group UK Ltd.
UKHW040102071024
449327UK00019B/746